# SpringerBriefs in Mathematical Physics

Volume 3

More information about this series at http://www.springer.com/series/11953

Marcel Bischoff · Yasuyuki Kawahigashi
Roberto Longo · Karl-Henning Rehren

# Tensor Categories and Endomorphisms of von Neumann Algebras

## with Applications to Quantum Field Theory

Springer

Marcel Bischoff
Institut für Theoretische Physik
Universität Göttingen
Göttingen
Germany

Yasuyuki Kawahigashi
Department of Mathematical Sciences
and Kavli IPMU (WPI)
The University of Tokyo
Tokyo
Japan

Roberto Longo
Dipartimento di Matematica
Università di Roma "Tor Vergata"
Rome
Italy

Karl-Henning Rehren
Institut für Theoretische Physik
Universität Göttingen
Göttingen
Germany

ISSN 2197-1757 ISSN 2197-1765 (electronic)
SpringerBriefs in Mathematical Physics
ISBN 978-3-319-14300-2 ISBN 978-3-319-14301-9 (eBook)
DOI 10.1007/978-3-319-14301-9

Library of Congress Control Number: 2014958015

Springer Cham Heidelberg New York Dordrecht London


Printed on acid-free paper

Springer International Publishing AG Switzerland is part of Springer Science+Business Media (www.springer.com)

# Preface

Subfactors (unital inclusions of von Neumann algebras with trivial centre) became a thriving focus of research interest after Vaughan Jones discovered in 1983 the quantization of the index below four. The associated principal graph was immediately identified as an important combinatorial invariant beyond the index, controlling the induction and restriction of bimodules of and between the two factors; more detailed information is encoded in the "planar algebra".

There is a close similarity with the theory of superselection sectors in relativistic quantum field theory (QFT), which was developed in the late 1960s and early 1970s by Sergio Doplicher, Rudolf Haag and John E. Roberts in Algebraic Quantum Field Theory (DHR). Especially, the method to obtain the quantization of the index closely resembled the argument for the quantization of the "statistical dimension" in the DHR theory. This involves (independent) works of two of us in 1989: R.L. found the direct link between the statistical dimension and the Jones index, and K.-H.R. (with K. Fredenhagen and B. Schroer) studied the braided tensor categorical superselection structure in low-dimensional quantum field theory.

While the details of the two theories differ (Jones theory addresses type II factors, whereas the local algebras of QFT are generically type III factors), the underlying mathematical structure is in both cases a C* tensor category (or 2-category). In the first case, its objects are bimodules, in the latter case they are endomorphisms (or homomorphisms); but as abstract structures, one deals with "the same" categories. A main purpose of the present work is in fact to "transfer" concepts from abstract tensor categories (going beyond just subfactor theory) into the language of von Neumann algebras and their endomorphisms, see below.

A main aim of the DHR theory in four spacetime dimensions (finally achieved in 1990 by S. Doplicher and J.E. Roberts) was to establish the identification of the category of DHR endomorphisms with a dual of a compact group, which is then the global gauge group of an extended quantum field theory (the "field algebra") containing the original QFT as its gauge fixed points. Crucial for this identification was the existence of a braiding, which is in fact maximally degenerate (i.e., a "permutation symmetry") in the DHR category.

This is markedly different in low-dimensional quantum field theory, notably in chiral and two-dimensional conformal QFT, which also experienced a research boost in the mid-1980s after the complementary breakthrough discoveries of A.A. Belavin, A.M. Polyakov, A.B. Zamolodchikov (minimal models) and D. Friedan, Z. Qiu, S. Shenker (classification of positive-energy representations of the Virasoro algebra). Again, the sectors of these theories are described by a braided C* tensor category, but the braiding turned out to be non-degenerate (modular) in most models of interest; a structural argument about why (and when) this is the case was given in 2001 by two of us (Y.K. and R.L.) in collaboration with M. Müger.

Not least for the reason that these structures had been discovered both in the physics context and in connection with quantum groups at about the same time, the focus of mathematical interest concentrated on modular tensor categories, which appear to describe generalized symmetries akin to group symmetries, but placed "at the other end of the range of possibilities" (tensor categories with modular braiding vs. tensor categories with symmetric braiding). Not only classification results were obtained, but relations with different fields of mathematics (vertex operator algebras, algebraic topology, elliptic functions) were discovered and explored.

On the physics side, the idea was put forward to hinge the axiomatic definition of a conformal QFT on its modular tensor category. While the present authors do not entirely conform to this idea (because it would exclude important models), it was certainly very fruitful for the discussion of a large class of interesting models.

A most important insight emerged from the formulation of "topological quantum field theory" (TQFT) in terms of the data of a given modular category, promoted by Jürgen Fuchs, Jürg Fröhlich, Ingo Runkel, Christoph Schweigert et al. (FFRS) in the late 1990s until today. This is the insight that the effect of representation-changing spacetime boundaries is entirely controlled by structures within the modular category: notably modules and bimodules of Frobenius algebras.

Frobenius algebras in a C* tensor category of endomorphisms had also been discovered—under the name of Q-systems—by one of us (R.L.) in 1994 as a complete invariant for type III subfactors $N \subset M$ of finite index. A crucial aspect is that the relevant category is the category of endomorphisms of the *smaller* factor $N$, so that the *larger* factor $M$ is characterized in terms of data pertaining to $N$. This changes the perspective from *subfactors* ("$N$ is embedded into a given $M$") to *extensions* ("$M$ extends a given $N$"). In fact, for single subfactors, there is a duality (related to the Jones tower) by which an extension $N \subset M$ is equivalently described by a subfactor $\gamma(M) \subset N$ (where $\gamma$ is a canonical endomorphism of $M$ with values in $N$), so that this change in perspective seems to be just a matter of taste. However, it becomes crucial in the application to QFT, where $N$ and $M$ have a direct physical meaning while $\gamma(M)$ has not.

It is in this field of research where the interests of the four of us have eventually converged.

Two of us had noticed the relevance of subfactor theory, and in particular the characterization of extensions in terms of Q-systems, for questions like, "Which quantum field theories possibly share the same stress-energy tensor" (or some other common sub-theory)? In chiral conformal QFT, the stress-energy tensor is described by the Virasoro algebra, whose positive-energy representations are known and give rise to well-studied modular C* tensor categories (provided the central charge is $c < 1$). Indeed, full classifications have been obtained along this line.

Especially, the formulation of boundaries and boundary conditions in relativistic conformal QFT, and its relation to the remarkable findings in the TQFT approach by FFRS, have intrigued us. The present work, along with several research papers, is an outcome of the endeavour to gain a better understanding of these connections.

A main difference with the TQFT approach is that a TQFT is essentially *defined* in terms of a modular tensor category (which need not be a C* tensor category), whereas in our feeling, a conformal QFT is in the first place a relativistic quantum field theory with an enhanced symmetry, subject to well-established axioms among which the C* structure (crucial for quantum observables) and local commutativity (Einstein causality) are most essential. In this vein, the presence of the modular category has to be *derived* (via the DHR theory), and its role in the formulation of QFT with boundaries has to be established.

In the present work, we focus on the theory of (modular) C* tensor categories, only keeping the applications to QFT in the back of our minds, and devoting the final chapter to a review of these applications. Large parts of the abstract theory were originally developed by FFRS, involving more recently also L. Kong; our main contribution is to clarify the "transfer" of these results into the language of endomorphisms of von Neumann algebras (which then facilitates the intended application to QFT).

## Acknowledgments

We are much indebted to J. Fuchs, I. Runkel, and C. Schweigert for their hospitality and enlightening explanations of their work, which were most beneficial for the results presented in Sect. 4.12. Y.K. thanks M. Izumi for an interesting question.

We also acknowledge institutional and financial support:

- Support by the Grants-in-Aid for Scientific Research, JSPS (Y.K.)
- Support by the German Research Foundation (Deutsche Forschungsgemeinschaft (DFG)) through the Institutional Strategy of the University of Göttingen (M.B., K.-H.R.)
- Support by the Alexander von Humboldt Foundation and the European Research Council (R.L.)

• Hospitality and support of the Erwin Schrödinger International Institute for Mathematical Physics, Vienna (all of us).

| | |
|---|---|
| Nashville, October 2014 | Marcel Bischoff |
| Tokyo | Yasuyuki Kawahigashi |
| Rome | Roberto Longo |
| Göttingen | Karl-Henning Rehren |

# Contents

# Chapter 1
# Introduction

**Abstract** Q-systems describe "extensions" of an infinite von Neumann factor $N$, i.e., finite-index unital inclusions of $N$ into another von Neumann algebra $M$. They are (special cases of) Frobenius algebras in the C* tensor category of endomorphisms of $N$. We review the relation between Q-systems, their modules and bimodules as structures in a tensor category on one side, and homomorphisms between von Neumann algebras on the other side. We then elaborate basic operations with Q-systems (various decompositions in the general case, and the centre, the full centre, and the braided product in braided categories), and illuminate their meaning in the von Neumann algebra setting. The main applications are in local quantum field theory, where Q-systems in the subcategory of DHR endomorphisms of a local algebra encode extensions $\mathscr{A}(O) \subset \mathscr{B}(O)$ of local nets. These applications, notably in conformal quantum field theories with boundaries, are briefly exposed, and are discussed in more detail in two original papers [1, 2].

Q-systems have first appeared in [3] as a device to characterize finite-index subfactors $N \subset M$ of infinite (type III) von Neumann algebras, generalizing the Jones theory of type II subfactors [4–6]. A Q-system is a triple

$$\mathbf{A} = (\theta, w, x),$$

where $\theta$ is a unital endomorphism of $N$ and $w \in \mathrm{Hom}(\mathrm{id}_N, \theta)$, $x \in \mathrm{Hom}(\theta, \theta^2)$ are a pair of intertwiners whose algebraic relations guarantee that $\theta$ is the dual canonical endomorphism (Sect. 2.2) associated with a subfactor $N \subset M$.

Notice that the data of the Q-system pertain only to $N$, so the Q-system actually characterizes $M$ as an "extension" of $N$. In fact, the larger algebra $M$ along with the embedding of $N$ into $M$ can be explicitly reconstructed (up to isomorphism) from the data. One issue in this work is a generalization to Q-systems for extensions $N \subset M$ where $M$ may have a finite centre, i.e., $M$ is a direct sum of infinite factors.

Subfactors are, apart from their obvious mathematical interest, also of physical interest since they describe, e.g., the embedding of a physical quantum sub-system in larger system. In this context, it is essential that the algebras are C* or von Neumann algebras, since quantum observables are always (selfadjoint) elements of such algebras, and in relativistic QFT, local observables generate factors of type III.



M. Bischoff et al., *Tensor Categories and Endomorphisms of von Neumann Algebras*, SpringerBriefs in Mathematical Physics, DOI 10.1007/978-3-319-14301-9_1

From a category point of view, a Q-system is the same as a special C* Frobenius algebra in a (strict, simple) C* tensor category. In the case at hand, the category would be (a subcategory of) the category $\mathrm{End}_0(N)$ of endomorphisms of $N$ with finite dimension. This is actually the most general situation, since every (rigid, countable) abstract C* tensor category can be realized as a full subcategory of $\mathrm{End}_0(N)$ [7].

In a more general setting (notably without assuming the C* structure which is naturally present in the case of $\mathrm{End}_0(N)$) abstract tensor categories and Frobenius algebras have been extensively studied by many mathematicians [8–12], and interesting "derived" structures have been discovered and classified, notably when the underlying tensor category is braided, or even modular [13–20].

A connection to physics of this more general setting is provided by [21–24] where a formulation of two-dimensional (*Euclidean*) conformal quantum field theory on Riemannian surfaces is developed in terms of a three-dimensional "topological quantum field theory" which is a cobordism theory between pairs of Riemannian surfaces. The authors observed, among a wealth of other results, that the modules and bimodules of the representation category of the underlying chiral theory play a prominent role in the classification of one-dimensional boundaries between Riemannian surfaces.

From the von Neumann algebra point of view, an important class of braided tensor subcategories of $\mathrm{End}_0(N)$ naturally arises in the algebraic formulation of *relativistic* Quantum Field Theory (QFT). Namely, a distinguished class of positive-energy representations of local QFT can be described in terms of endomorphisms of the C* algebra $\mathscr{A}$ of quasi-local observables. These DHR endomorphisms are the objects of a braided C* tensor category [25, 26]. By restricting attention to a von Neumann algebra $N = \mathscr{A}(O)$ of local observables, one obtains a braided tensor subcategory of $\mathrm{End}_0(N)$. In this context, Q-systems describe finite-index extensions $\mathscr{A} \subset \mathscr{B}$ of quantum field theories, and $\mathscr{B}$ is local if and only if the Q-system is commutative w.r.t. the braiding.

Our main motivation for the present work was the study of boundary conditions in relativistic conformal QFT in two spacetime dimensions, as discussed in detail in the compagnon papers [1, 2]. Boundaries in relativistic quantum field theories [27–29], with observables that are Hilbert space operators subject to the principle of locality (or rather causality), have been analyzed much less than in the Euclidean setting. Very little is known about an apriori relation between Euclidean and Lorentzian boundaries. Yet, our treatment of boundaries in relativistic two-dimensional conformal QFT shows that precisely the same mathematical structures, namely the chiral representation category, its Q-systems and their modules and bimodules, control the boundary conditions in both situations. We address in particular the case of "hard" boundaries in [1] and "transparent" or "phase boundaries" (defects) in [2]. In this work, we shall concentrate on the underlying mathematical theory, with only scattered remarks about the relevance in QFT. A brief exposition of these physical applications will be given in Chap. 5.

While large portions of the category side of this work are reformulations from [30, 31], our original contribution is the elaboration of the relation between the abstract category notions and the von Neumann algebra setting and subfactor theory. A prominent issue is our proof of Theorem 4.42 (a characterization of the central projections of an extension $N \subset M$, which is given by the braided product of two full-centre Q-systems in a modular category). This theorem is implicitly present, but widely scattered in the work of [21–24, 30–33]. Our proof is much more streamlined, because it benefits from substantial simplifications in the C* setting, where one can exploit positivity arguments in crucial steps.

This theorem is relevant for phase boundaries in relativistic two-dimensional conformal QFT because it classifies the boundary conditions in terms of chiral data [2], very much the same as in the Euclidean setting [21].

Other original contributions in this work concern Q-systems for extensions $N \subset M$ when $M$ is not a factor, a situation that naturally occurs in several applications, as well as the characterization of various types of decompositions of Q-systems (Sects. 4.2–4.4) in terms of algebraic properties of projections in $\mathrm{Hom}(\theta, \theta)$.

In Chap. 2, we review the basic notions concerning endomorphisms and homomorphisms of infinite von Neumann algebras, with special emphasis on the notions of conjugates and dimension.

Chapter 3 is devoted to the category structure, and to the correspondences between Q-systems and algebra extensions, and between bimodules between Q-systems and homomorphisms between the corresponding extensions.

Chapter 4 is the main part of this work. We introduce various operations with Q-systems (decompositions, braided products, centres and full centre), and investigate their meaning in the setting of von Neumann algebras.

Chapter 5 contains an exposition of the appearance of braided and modular C* tensor categories in the DHR theory of superselection sectors in Algebraic QFT, and reviews the relevance of Q-systems for issues like extensions and boundary conditions.

## References


1. M. Bischoff, Y. Kawahigashi, R. Longo, Characterization of 2D rational local conformal nets and its boundary conditions: the maximal case. arXiv:1410.8848
2. M. Bischoff, Y. Kawahigashi, R. Longo, K.-H. Rehren, Phase boundaries in algebraic conformal QFT. arXiv:1405.7863
3. R. Longo, A duality for Hopf algebras and for subfactors. Commun. Math. Phys. **159**, 133–150 (1994)
4. V.F.R. Jones, Index for subfactors. Invent. Math. **72**, 1–25 (1983)
5. H. Kosaki, Extension of Jones' theory on index to arbitrary factors. J. Funct. Anal. **66**, 123–140 (1986)
6. S. Popa, *Classification of Subfactors and Their Endomorphisms*, CBMS Regional Conference Series in Mathematics, vol. 86 (American Mathematical Society, Providence, 1995)
7. S. Yamagami, C*-tensor categories and free product bimodules. J. Funct. Anal. **197**, 323–346 (2003)
8. P. Deligne, Catégories tannakiennes, in *Grothendieck Festschrift*, vol. II, ed. by P. Cartier, et al. (Birkhäuser, Basel, 1991), pp. 111–195
9. A. Joyal, R. Street, Braided tensor categories. Adv. Math. **102**, 20–78 (1993)
10. S. Mac Lane, *Categories for the Working Mathematician*, 2nd edn. (Springer, Berlin, 1998)

11. B. Bakalov, A. Kirillov Jr., *Lectures on Tensor Categories and Modular Functors*, University Lecture Series, vol. 21 (AMS, Providence, 2001)
12. P. Deligne, Catégories tensorielles. Mosc. Math. J. **2**, 227–248 (2002)
13. M. Müger, Galois theory for braided tensor categories and the modular closure. Adv. Math. **150**, 151–201 (2000)
14. A. Kirillov Jr., V. Ostrik, On $q$-analog of McKay correspondence and ADE classification of $sl(2)$ conformal field theories. Adv. Math. **171**, 183–227 (2002)
15. V. Ostrik, Module categories, weak Hopf algebras and modular invariants. Transform. Groups **8**, 177–206 (2003)
16. M. Müger, From subfactors to categories and topology I. Frobenius algebras in and Morita equivalence of tensor categories. J. Pure Appl. Algebra **180**, 81–157 (2003)
17. M. Müger, From subfactors to categories and topology II. The quantum double of tensor categories and subfactors. J. Pure Appl. Algebra **180**, 159–219 (2003)
18. M. Müger, Galois extensions of braided tensor categories and braided crossed G-categories. J. Algebra **277**, 256–281 (2004)
19. P. Etingof, D. Nikshych, V. Ostrik, On fusion categories. Ann. Math. **162**, 581–642 (2005)
20. A. Davydov, M. Müger, D. Nikshych, V. Ostrik, The Witt group of non-degenerate braided fusion categories. arXiv:1009.2117
21. J. Fuchs, I. Runkel, C. Schweigert, TFT construction of RCFT correlators I: partition functions. Nucl. Phys. B **646**, 353–497 (2002)
22. J. Fuchs, I. Runkel, C. Schweigert, TFT construction of RCFT correlators II. Nucl. Phys. B **678**, 511–637 (2004)
23. J. Fuchs, I. Runkel, C. Schweigert, TFT construction of RCFT correlators III. Nucl. Phys. B **694**, 277–353 (2004)
24. J. Fuchs, I. Runkel, C. Schweigert, TFT construction of RCFT correlators IV. Nucl. Phys. B **715**, 539–638 (2005)
25. S. Doplicher, R. Haag, J.E. Roberts, Local observables and particle statistics. I. Commun. Math. Phys. **23**, 199–230 (1971)
26. K. Fredenhagen, K.-H. Rehren, B. Schroer, Superselection sectors with braid group statistics and exchange algebras I. Commun. Math. Phys. **125**, 201–226 (1989)
27. R. Longo, K.-H. Rehren, Local fields in boundary CFT. Rev. Math. Phys. **16**, 909–960 (2004)
28. R. Longo, K.-H. Rehren, How to remove the boundary in CFT—an operator algebraic procedure. Commun. Math. Phys. **285**, 1165–1182 (2009)
29. S. Carpi, Y. Kawahigashi, R. Longo, How to add a boundary condition. Commun. Math. Phys. **322**, 149–166 (2013)
30. J. Fröhlich, J. Fuchs, I. Runkel, C. Schweigert, Correspondences of Ribbon categories. Ann. Math. **199**, 192–329 (2006)
31. L. Kong, I. Runkel, Morita classes of algebras in modular tensor categories. Adv. Math. **219**, 1548–1576 (2008)
32. J. Fröhlich, J. Fuchs, I. Runkel, C. Schweigert, Kramers-Wannier duality from conformal defects. Phys. Rev. Lett. **93**, 070601 (2004)
33. J. Fröhlich, J. Fuchs, I. Runkel, C. Schweigert, Duality and defects in rational conformal field theory. Nucl. Phys. B **763**, 354–430 (2007)

# Chapter 2
# Homomorphisms of von Neumann Algebras

**Abstract** We introduce the tensor category structure of endomorphisms of infinite (type III) von Neumann factors. We review the basic concepts of conjugate homomorphisms between a pair of infinite factors, including the dimension, and discuss the generalization to homomorphisms of a factor into a von Neumann algebra with a centre.

Let $N$ and $M$ be two von Neumann algebras, and $\alpha$, $\beta$ a pair of homomorphisms : $N \to M$. (Without further mentioning, the notion "homomorphism" will include the * and unit-preserving properties $\alpha(n^*) = \alpha(n)^*$ and $\alpha(\mathbf{1}_N) = \mathbf{1}_M$.) An operator $t \in M$ such that

$$t \cdot \alpha(n) = \beta(n) \cdot t \qquad \text{for all} \quad n \in N$$

is called an **intertwiner**, writing $t : \alpha \to \beta$ or $t \in \mathrm{Hom}(\alpha, \beta)$. Clearly, if $t \in \mathrm{Hom}(\alpha, \beta)$, then $t^* \in \mathrm{Hom}(\beta, \alpha)$; $\mathrm{Hom}(\alpha, \beta)$ is a complex vector space, and $\mathrm{Hom}(\alpha, \alpha)$ is a C*-algebra.

A homomorphism $\alpha : N \to M$ is composed with a homomorphism $\beta : M \to L$, such that $\beta \circ \alpha : N \to L$.

Likewise, for any three homomorphisms $\alpha, \beta, \gamma : N \to M$ and intertwiners $t \in \mathrm{Hom}(\alpha, \beta)$ and $s \in \mathrm{Hom}(\beta, \gamma)$, the product in $M$ gives an intertwiner $s \cdot t \in \mathrm{Hom}(\alpha, \gamma)$.

These structures turn the endomorphisms of a von Neumann algebra $N$ into a strict tensor category $\mathrm{End}(N)$, and the homomorphisms between von Neumann algebras $N, M, \ldots$ into a strict tensor 2-category, where the concatenation of morphisms is the product of intertwiners: $s \circ t := s \cdot t$, the monoidal product of objects is the composition of endomorphisms: $\beta \times \alpha := \beta \circ \alpha$, and the monoidal product of morphisms $t_i : \alpha_i \to \beta_i$ is the product

$$t_1 \times t_2 = t_1 \cdot \alpha_1(t_2) = \beta_1(t_2) \cdot t_1 :$$

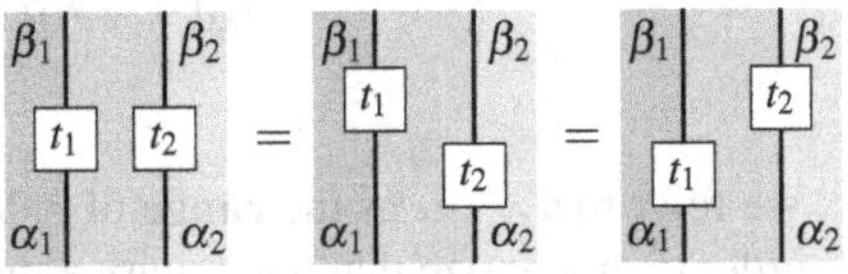


(This graphical notation, directly appealing to the underlying tensor category point of view, will render the structure of many algebraic computations more transparent.



M. Bischoff et al., *Tensor Categories and Endomorphisms of von Neumann Algebras*, SpringerBriefs in Mathematical Physics, DOI 10.1007/978-3-319-14301-9_2

Its basic rules are self-explaining from this example: Different shades indicate different von Neumann algebras, and we usually reserve the lightest shade for $N$, lines are homomorphisms, boxes and similar symbols to appear later are intertwiners, the monoidal product is horizontal juxtaposition, and the concatenation product is read from the bottom to the top. The operator adjoint is represented by up-down reflection.)

Notice that *as operators*, $t \times 1_\alpha = t$ is the same operator in a different intertwiner space, whereas $1_\alpha \times t = \alpha(t)$. To enhance readability, we shall occasionally suppress the concatenation symbol and write simply $s \circ t$ as the operator product $st$.

Because all intertwiner spaces $\mathrm{Hom}(\alpha, \beta)$ are linear subspaces of the target von Neumann algebra, they inherit its weak and norm topologies. In particular, $\mathrm{End}(N)$ is a C* tensor category, and the self-intertwiners $\mathrm{Hom}(\alpha, \alpha)$ form a C* algebra. Important consequences are that $t^* \circ t \equiv t^*t$ is a positive operator in $\mathrm{Hom}(\beta, \beta)$, and that $t^* \circ t = 0$ implies $t = 0$.

## 2.1 Endomorphisms of Infinite Factors

A von Neumann algebra $N$ is a **factor** iff its centre $N' \cap N \equiv \mathrm{Hom}(\mathrm{id}_N, \mathrm{id}_N) = \mathbb{C} \cdot \mathbf{1}_N$. Since $\mathrm{id}_N$ is the monoidal unit in the tensor category, this is the same as saying that the category $\mathrm{End}(N)$ is simple.

These elementary facts can be supplemented by further structure. If $u : \alpha \to \beta$ is unitary, $\alpha$ and $\beta$ are said to be **unitarily equivalent**. The unitary equivalence class of $\alpha$ is called the **sector** $[\alpha]$. An endomorphism $\alpha$ is **irreducible** iff $\mathrm{Hom}(\alpha, \alpha) = \mathbb{C} \cdot \mathbf{1}_N$.

In an **infinite** ($\Leftrightarrow$ purely infinite, type III) von Neumann factor acting on a separable Hilbert space (which we shall henceforth assume throughout), every projection $e \neq 0$ can be written as $e = ss^*$ where $s^*s = 1$, and one can always choose decompositions of the unit $1 = \sum_i s_i s_i^*$ such that $s_i^* s_j = \delta_{ij}$. The algebra generated by bounded quantum mechanical observables (= the algebra $\mathscr{B}(\mathscr{H})$ of all bounded operators) does not share this property; instead, the local algebras of quantum field theory are generically infinite von Neumann factors.

Thanks to this property, one can define

(i) an inclusion relation for endomorphisms: $\beta \prec \alpha$ iff there is $s : \beta \to \alpha$ with $s^*s = 1_\beta$.

(ii) subobjects: if $e : \alpha \to \alpha$ is a projection, then there is a sub-endomorphism $\alpha_s$ defined by the choice of $s$ such that $ss^* = e$, $s^*s = 1$, and putting

$$\alpha_s(\cdot) = s^*\alpha(\cdot)s :$$

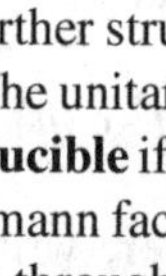


.

We refer to $\alpha_s \prec \alpha$ as the **range** of $e$. We shall sometimes write $\alpha_e$ instead, in order to emphasize that the unitary equivalence class of $\alpha_s$ does not depend on the choice of $s$. (Categories where subobjects exist are also called "Karoubian", thus $\mathrm{End}(N)$ is Karoubian if $N$ is an infinite factor.)

(iii) direct sums of endomorphisms:

$$\alpha(\cdot) := \sum_i s_i \alpha_i(\cdot) s_i^* : \qquad \sum_i$$

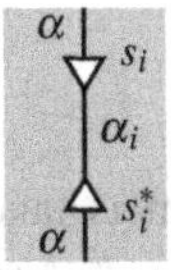


is an endomorphism, $\alpha_i \prec \alpha$. Suppressing the dependence on the isometries $s_i$, we write sloppily $\alpha \simeq \bigoplus_i \alpha_i$. Since the choice of the isometries $s_i$ is irrelevant for the unitary equivalence class (sector) $[\alpha]$, the direct sum should be understood as a direct sum of sectors. We emphasize this by writing also

$$[\alpha] = \bigoplus_i [\alpha_i].$$

## 2.2 Homomorphisms and Subfactors

All notions of the preceding presentation can be transferred to homomorphisms $\varphi : N \to M$ where both $N$ and $M$ are infinite factors. Notice that intertwiners $t \in \mathrm{Hom}(\varphi_1, \varphi_2)$ are elements of $M$.

Admitting several factors, one obtains a 2-category, whose objects are the factors, the 1-morphisms are the homomorphisms, and the 2-morphisms are their intertwiners.

If $N \subset M$ is a **subfactor** (i.e., both $N$ and $M$ are factors), then the identical map $\iota : N \to M, n \mapsto n$, is a nontrivial homomorphism, that describes the embedding of $N$ into $M$.

One can define [1, Chap. 3] a **dimension** function on the homomorphisms $N \to M$ when both $N$ and $M$ are infinite factors, which is additive under direct sums and multiplicative under composition. It is defined through the notion of **conjugates**: $\alpha : N \to M$ and $\overline{\alpha} : M \to N$ are said to be conjugates of each other whenever there is a pair of intertwiners $N \ni w : \mathrm{id}_N \to \overline{\alpha}\alpha$ and $M \ni \overline{w} : \mathrm{id}_M \to \alpha\overline{\alpha}$ satisfying the **conjugacy relations**

$$(w^* \times 1_{\overline{\alpha}}) \circ (1_{\overline{\alpha}} \times \overline{w}) = 1_{\overline{\alpha}} :$$

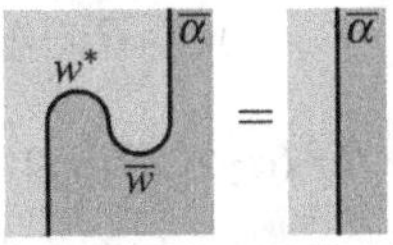


$$(1_\alpha \times w^*) \circ (\overline{w} \times 1_\alpha) = 1_\alpha :$$

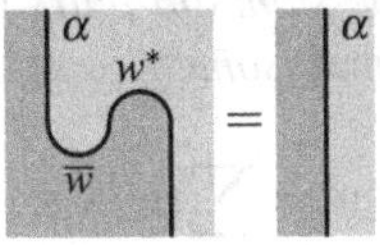


(2.2.1)

Being self-intertwiners of $\mathrm{id}_N$, resp. $\mathrm{id}_M$, $w^*w = d \cdot \mathbf{1}_N$ and $\overline{w}^*\overline{w} = d' \cdot \mathbf{1}_M$ are positive scalars, and $w, \overline{w}$ can be normalized such that $d = d'$. The **dimension**

$\dim(\alpha) = \dim(\overline{\alpha})$ is defined to be

$$\dim(\alpha) = \dim(\overline{\alpha}) := \inf_{(w,\overline{w})} d \tag{2.2.2}$$

where the infimum is taken over all solutions $(w, \overline{w})$ of the conjugacy relations Eq. (2.2.1) with $d = d'$. A solution saturating the infimum is called **standard solution** or **standard pair**. If $\alpha$ and $\beta$ are irreducible, every solution with $d = d'$ is standard, because $\dim\mathrm{Hom}(\mathrm{id}, \alpha\overline{\alpha}) = \dim\mathrm{Hom}(\overline{\alpha}\alpha)$=1. In the general case, standard solutions always exists, and are unique up to unitary equivalence [1, 2].

(Here is a simple explicit proof: For $[\alpha] = \bigoplus_i n_i[\alpha_i]$ and $[\overline{\alpha}] = \bigoplus_i \overline{n}_i[\overline{\alpha}_i]$ with $\alpha_i, \overline{\alpha}_i$ irreducible, one may choose standard pairs $(w_i, \overline{w}_i)$ for $\alpha_i, \overline{\alpha}_i$ and orthonormal bases $s_a^i \in \mathrm{Hom}(\alpha_i, \alpha)$, $\overline{s}_b^i \in \mathrm{Hom}(\overline{\alpha}_i, \overline{\alpha})$. Then the most general element of $\mathrm{Hom}(\mathrm{id}, \overline{\alpha}, \alpha)$ is of the form $w = \sum_i \sum_{ab} c_{ab}^i \overline{\alpha}(s_a^i)\overline{s}_b^i w_i$, and similarly $\overline{w} = \sum_i \sum_{ab} c'^i_{ab}\, \alpha(\overline{s}_b^i) s_a^i \overline{w}_i$. These solve the conjugacy relations iff the coefficient matrices satisfy $c'^i = (c^i)^{-1*}$ (in particular, the multiplicities $\overline{n}_i = n_i$ must be the same), and one has $d = \sum_i \dim(\alpha_i)\,\mathrm{Tr}(c^i)^* c^i$, $d' = \sum_i \dim(\overline{\alpha}_i)\,\mathrm{Tr}(c^i)^{-1*}(c^i)^{-1}$. The variational problem $d[c]d'[c] \overset{!}{=} \min$ with $d = d'$ is solved by any family of unitary matrices $c^i$.)

The conjugate of an endomorphism is unique up to unitary equivalence. Endomorphisms which do not have conjugates can be assigned the dimension $\infty$.

The dimension is always $\geq$1, and a homomorphism $\alpha$ is an isomorphism iff $\dim(\alpha) = 1$. In this case, $\alpha^{-1}$ is a conjugate of $\alpha$. More generally, the dimension is the square root of the (minimal) index [3, 4]:

$$\dim(\alpha)^2 = [M : \alpha(N)].$$

In particular, for a subfactor $N \subset M$, $\dim(\iota)$ is the square root of the index $[M : N]$ [5]. In this case, $\iota\overline{\iota} \in \mathrm{End}(M)$ is called the **canonical endomorphism**, and $\overline{\iota}\iota \in \mathrm{End}(N)$ the **dual canonical endomorphism**.

**Lemma 2.1** ([1]) (i) *Let* $(w_1, \overline{w}_1)$ *and* $(w_2, \overline{w}_2)$ *be standard pairs for* $(\alpha_1, \overline{\alpha}_1)$ *and for* $(\alpha_2, \overline{\alpha}_2)$*, respectively. Then*

$$w = \overline{\alpha}_1(w_2)w_1, \quad \overline{w} = \alpha_2(\overline{w}_1)\overline{w}_2$$

*is a standard pair for* $(\alpha_2\alpha_1, \overline{\alpha}_1\overline{\alpha}_2)$.

(ii) *Let* $(w_i, \overline{w}_i)$ *be standard pairs for* $(\alpha_i, \overline{\alpha}_i)$*, and* $[\alpha] = \bigoplus_i[\alpha_i]$, $[\overline{\alpha}] = \bigoplus_i[\overline{\alpha}_i]$. *Choose orthonormal isometries* $s_i \in \mathrm{Hom}(\alpha_i, \alpha)$ *and* $\overline{s}_i \in \mathrm{Hom}(\overline{\alpha}_i, \overline{\alpha})$. *Then*

$$w = \sum_i (\overline{s}_i \times s_i) \circ w_i, \quad \overline{w} = \sum_i (s_i \times \overline{s}_i) \circ \overline{w}_i$$

*is a standard pair for* $(\alpha, \overline{\alpha})$.

**Corollary 2.2** ([1]) *The conjugate respects direct sums, and the dimension is additive and multiplicative:*

$$\dim(\alpha_2 \circ \alpha_1) = \dim(\alpha_2) \cdot \dim(\alpha_1), \quad \dim(\alpha) = \sum_i \dim(\alpha_i) \ \text{ if } [\alpha] = \bigoplus_i [\alpha_i].$$

It should, of course, be emphasized that all the notions of direct sums, subobjects, conjugates and dimension respect the unitary equivalence relation.

**Definition 2.3** The left and right **traces** are the faithful positive maps

$$\mathrm{LTr}_\alpha : \mathrm{Hom}(\alpha\beta, \alpha\beta') \to \mathrm{Hom}(\beta, \beta'),$$
$$t \mapsto (w^* \times 1_\beta) \circ (1_{\overline{\alpha}} \times t) \circ (w \times 1_\beta)$$

(2.2.3)

$$\mathrm{RTr}_\alpha : \mathrm{Hom}(\beta\alpha, \beta'\alpha) \to \mathrm{Hom}(\beta, \beta'),$$
$$t \mapsto (1_\beta \times \overline{w}^*) \circ (t \times 1_{\overline{\alpha}}) \circ (1_\beta \times \overline{w})$$

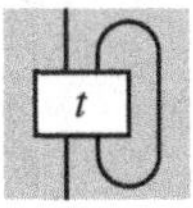


(2.2.4)

for the conjugate homomorphisms $\alpha, \overline{\alpha}$.

**Proposition 2.4** ([1, Lemma 3.7]) *Let $N$ and $M$ be infinite factors, and let the traces* $\mathrm{LTr}_\alpha$ *and* $\mathrm{RTr}_\alpha$ *be defined w.r.t. a standard solution $(w, \overline{w})$ of the conjugacy relations for $\alpha : N \to M$ and $\overline{\alpha} : M \to N$.*

*The traces do not depend on the choice of the conjugate and of the standard solution, and satisfy the trace property*

$$\mathrm{LTr}_\alpha(s \times 1_{\beta'}) \circ t = \mathrm{LTr}_{\alpha'} t \circ (s \times 1_\beta)$$

=

,

$$\mathrm{RTr}_\alpha(1_{\beta'} \times s) \circ t = \mathrm{RTr}_{\alpha'} t \circ (1_\beta \times s)$$

=

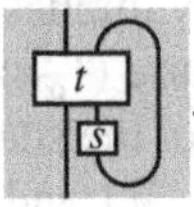


.

(2.2.5)

*for $s \in \mathrm{Hom}(\alpha', \alpha)$ and $t \in \mathrm{Hom}(\alpha\beta, \alpha'\beta')$ resp. $t \in \mathrm{Hom}(\beta\alpha, \beta'\alpha')$. For $\beta = \beta' = \mathrm{id}$, both traces coincide and are denoted* $\mathrm{Tr}_\alpha : \mathrm{Hom}(\alpha, \alpha) \to \mathbb{C}$. *In particular,*

$$\mathrm{Tr}_\alpha 1_\alpha = \dim(\alpha). \tag{2.2.6}$$

The latter property can in fact be adopted as an alternative definition for standardness, since one also has

**Proposition 2.5** ([1, Lemma 3.9]) *Let $N$ and $M$ be infinite factors, and let the traces* $\mathrm{LTr}_\alpha$ *and* $\mathrm{RTr}_\alpha$ *be defined as in Definition* 2.3 *w.r.t.* any (*i.e., not necessarily standard*)

*solution* $(w, \overline{w})$ *of the conjugacy relations for* $\alpha : N \to M$ *and* $\overline{\alpha} : M \to N$. *Then* $\mathrm{LTr}_\alpha$ *and* $\mathrm{RTr}_\alpha$ *coincide if and only if* $(w, \overline{w})$ *is standard.*

If $(w, \overline{w})$ is not standard, the maps $\mathrm{LTr}_\alpha$ and $\mathrm{RTr}_\alpha$ on $\mathrm{Hom}(\alpha, \alpha) \to \mathbb{C}$ may happen to be traces, without being equal. E.g., for reducible $\alpha$ every $n \in \mathrm{Hom}(\alpha, \alpha)$ gives rise to a deformation $w' := (1_{\overline{\alpha}} \times n) \circ w$, $\overline{w}' := (n^{*-1} \times 1_{\overline{\alpha}}) \circ \overline{w}$ of a standard pair $(w, \overline{w})$, which still solves the conjugacy relations. Then $\mathrm{LTr}'_\alpha$ and $\mathrm{RTr}'_\alpha$ defined with $(w', \overline{w}')$ are traces if and only if $n^*n$ is central in $\mathrm{Hom}(\alpha, \alpha)$, while $(w', \overline{w}')$ is standard iff $n^*n = 1_\alpha$. One has the following characterization [1, Lemma 2.3]:

**Proposition 2.6** *Let* $(w, \overline{w})$ *and* $(w', \overline{w}')$ *be solutions of the conjugacy relations for* $\alpha, \overline{\alpha}$ *and for* $\alpha', \overline{\alpha}'$, *not necessarily standard. Define* $\mathrm{LTr}_\alpha$ *as in Definition* 2.3 *w.r.t. these pairs. The following are equivalent:*

(i) *For* $t \in \mathrm{Hom}(\alpha, \alpha')$ *and* $s \in \mathrm{Hom}(\alpha', \alpha)$, *one has* $\mathrm{LTr}_\alpha(st) = \mathrm{LTr}_{\alpha'}(ts)$.
(ii) *For* $t \in \mathrm{Hom}(\alpha, \alpha')$, *one has*

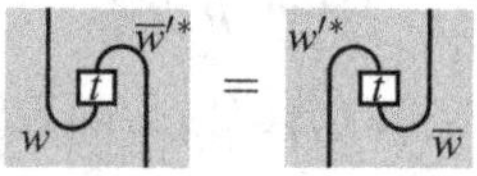

$\in \mathrm{Hom}(\overline{\alpha}, \overline{\alpha}')$.

*The same is true, replacing* LTr *by* RTr *in* (i), *or replacing* $t$ *by* $s \in \mathrm{Hom}(\alpha', \alpha)$ *in* (ii).

*In particular,* (ii) *holds if* $(w, \overline{w})$ *and* $(w', \overline{w}')$ *are standard.*

*Proof* "(i) ⇒ (ii)" is the statement of [1, Lemma 2.3c], although the authors actually prove also the converse. The proof proceeds by noting that

$\mathrm{LTr}_\alpha(st) =$ 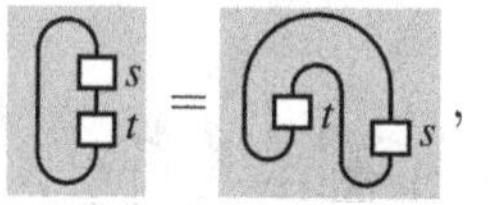

, $\mathrm{RTr}_{\alpha'}(ts) =$ 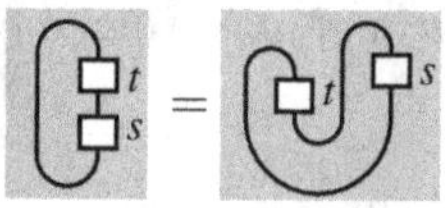

.

Now, (ii) trivially implies equality of the two expressions, hence (i). Conversely, (i) implies (ii) because $(1_{\overline{\alpha}'} \times s) \circ w'$ is an arbitrary element of $\mathrm{Hom}(\mathrm{id}, \overline{\alpha}\alpha')$.

The variants of the statement follow by obvious modifications.

Finally, if $(w, \overline{w})$ and $(w', \overline{w}')$ are standard, then Proposition 2.4 implies (i), hence (ii). □

For a single infinite von Neumann factor $N$, $\mathrm{End}_0(N)$ is the full subcategory of $\mathrm{End}(N)$, whose objects are the endomorphisms of finite dimension. This is a "rigid" category since left and right duals exist for all objects (namely, the conjugate).

All intertwiner spaces $\mathrm{Hom}(\alpha, \beta)$ in $\mathrm{End}_0(N)$ are finite-dimensional, and $\mathrm{Hom}(\alpha, \alpha)$ are isomorphic with a direct sum of matrix algebras $\bigoplus_\lambda \mathrm{Mat}_{\mathbb{C}}(n_\lambda)$, where $\lambda$ are the equivalence classes of irreducible sub-endomorphisms of $\alpha$ and $n_\lambda$ their multiplicities in $\alpha$.

Whenever $\alpha$ has finite dimension (and hence a conjugate $\overline{\alpha}$ exists), one can use a standard solution $(w, \overline{w})$ to define linear bijections (left and right **Frobenius conjugations**) between the spaces $\mathrm{Hom}(\gamma_2, \alpha\gamma_1)$ and $\mathrm{Hom}(\overline{\alpha}\gamma_2, \gamma_1)$, and between $\mathrm{Hom}(\gamma_2, \gamma_1\alpha)$ and $\mathrm{Hom}(\gamma_2\overline{\alpha}, \gamma_1)$,

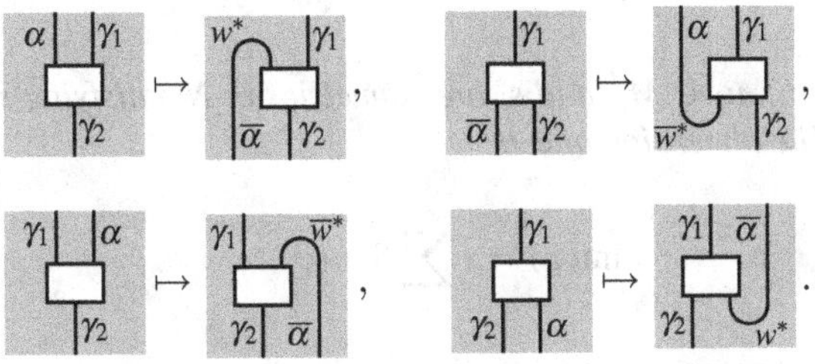


These maps along with the ensuing equalities of the dimensions of the intertwiner spaces,

$$\dim\mathrm{Hom}(\gamma_2, \alpha\gamma_1) = \dim\mathrm{Hom}(\overline{\alpha}\gamma_2, \gamma_1),$$
$$\dim\mathrm{Hom}(\gamma_2, \gamma_1\alpha) = \dim\mathrm{Hom}(\gamma_2\overline{\alpha}, \gamma_1),$$

are usually referred to as **Frobenius reciprocities**.

## 2.3 Non-factorial Extensions

We want to extend our setup to $N$ being a factor, while $M$ is admitted to be a properly infinite von Neumann algebra with finite centre. For a related analysis, see [6, 7].

$M$ is a direct sum of finitely many infinite factors

$$M = \bigoplus_i M_i.$$

The units of $M_i$ are the minimal central projections $e_i$ of $M$. A homomorphism $\varphi : N \to M$ can then be written as

$$\varphi(n) = \bigoplus_i \varphi_i(n).$$

Unlike the direct sum of sectors involving isometric intertwiners, cf. Sect. 2.1, this is the true direct sum of homomorphisms $\varphi_i : N \to M_i$, which is a homomorphism $N \to \bigoplus_i M_i$.

Notice that the central projections $e_i \in M$ are self-intertwiners of $\varphi$, but $e_i$ can *not* be split as $ss^*$ with isometries $s \in M$. Therefore, the direct sum of sectors $[\varphi_i]$ as in Sect. 2.1 is not defined.

**Proposition 2.7** *If all $\varphi_i : N \to M_i$ have conjugates $\overline{\varphi}_i$, then a conjugate homomorphism $\overline{\varphi} : M \to N$ of $\varphi$ can be defined as*

$$\overline{\varphi}(m) = \sum_i s_i \overline{\varphi}_i(m_i) s_i^*$$

*where $m = \bigoplus_i m_i$, $m_i \in M_i$, and $s_i$ are isometries in $N$ satisfying $s_i^* s_j = \delta_{ij}$ and $\sum_i s_i s_i^* = \mathbf{1}_N$. The dimension of $\varphi$ is*

$$\dim(\varphi) = \Big(\sum_i \dim(\varphi_i)^2\Big)^{\frac{1}{2}}. \tag{2.3.1}$$

The dimension $\dim(\varphi)$ is defined by the same infimum as Eq. (2.2.2), taken over all solutions $(w, \overline{w})$ of the conjugacy relations such that $w^* w = d \cdot \mathbf{1}_N, \overline{w}^* \overline{w} = d \cdot \mathbf{1}_M$. Notice that it is no longer additive, as in the factor case.

*Proof* One easily sees that the solutions of the conjugacy relations are parameterized by

$$w = \sum_i \lambda_i \cdot s_i w_i, \qquad \overline{w} = \bigoplus_i \overline{\lambda}_i^{-1} \cdot \varphi_i(s_i) \overline{w}_i,$$

with parameters $\lambda_i \in \mathbb{C}$. Here, $(w_i, \overline{w}_i)$ are solutions for $(\varphi, \overline{\varphi})$ satisfying $w_i^* w_i = d_i \cdot \mathbf{1}_N$ and $\overline{w}_i^* \overline{w}_i = d_i \cdot \mathbf{1}_{M_i}$. Imposing $w^* w = d \cdot \mathbf{1}_N$ and $\overline{w}^* \overline{w} = d \cdot \mathbf{1}_M$ fixes the numerical coefficients by $|\lambda_i|^2 = d/d_i$ and $d^2 = \sum_i d_i^2$. This quantity is minimized if all $d_i$ are minimal, i.e., all $(w_i, \overline{w}_i)$ are standard, and $d_i = \dim(\varphi_i)$. This completes the proof. □

*Remark 2.8* For standard pairs $(w, \overline{w})$ of multiples of isometries satisfying the minimality condition, the tracial properties (Propositions 2.4–2.6) fail in general, when $M$ (or $N$) is not a factor. The authors of [7] propose a different "normalization condition" (Eq. (4.3) in [7]) for solutions to the conjugacy relations, with $w^* w \in N$ and $\overline{w}^* \overline{w} \in M$ central but in general not multiples of $\mathbf{1}$. In the case of $N$ and $M$ both being factors, their condition amounts to the equality of the left and right traces, hence is equivalent to standardness by Proposition 2.5, but it distinguishes different normalizations otherwise. In the case at hand, it would rather fix $|\lambda_i|^2 = 1$, so that $\overline{w}^* \overline{w}$ is no longer a multiple of an isometry.

## References

1. R. Longo, J.E. Roberts, A theory of dimension. K-Theory **11**, 103–159 (1997). (notably Chaps. 3 and 4)
2. H. Kosaki, R. Longo, A remark on the minimal index of subfactors. J. Funct. Anal. **107**, 458–470 (1992)
3. R. Longo, Index of subfactors and statistics of quantum fields I. Commun. Math. Phys. **126**, 217–247 (1989)

4. R. Longo, A duality for Hopf algebras and for subfactors. Commun. Math. Phys. **159**, 133–150 (1994)
5. V.F.R. Jones, Index for subfactors. Invent. Math. **72**, 1–25 (1983)
6. F. Fidaleo, T. Isola, Minimal expectations for inclusions with atomic centre. Int. J. Math. Phys. **7**, 307–327 (1996)
7. A. Bartels, C.L. Douglas, A. Henriques, Dualizability and index of subfactors. arXiv:1110.5671

# Chapter 3
# Frobenius Algebras, Q-Systems and Modules

**Abstract** We introduce the notion of Q-systems as Frobenius algebras in a C* tensor category, enjoying a standardness property. Q-systems in the category of endomorphisms of an infinite factor $N$ completely characterize extensions $N \subset M$. Modules and bimodules of Q-systems are equivalent to homomorphisms $N \to M$ resp. $M_1 \to M_2$.

We collect here some relevant results about the (simple, strict, Karoubian) C* tensor category $\mathrm{End}_0(N)$ for an infinite von Neumann factor $N$. In fact, every full subcategory of $\mathrm{End}_0(N)$ can be canonically completed so as to become a simple strict Karoubian C* tensor category with direct sums

$$\mathscr{C} \subset \mathrm{End}_0(N).$$

This completion is precisely given by the constructions exposed in Sect. 2.1. Without further specification, throughout this work $\mathscr{C} \subset \mathrm{End}_0(N)$ will denote a subcategory with the stated properties.

In the motivating application to QFT, as exposed in Chap. 5.1.2, $N$ will be the von Neumann algebra $\mathscr{A}(O)$ of observables localized in some region $O$ of spacetime, which is known to be an infinite factor under very general assumptions. The assignment $O \mapsto \mathscr{A}(O)$ is called the local net of observables, and a distinguished class of positive-energy representations can be described by DHR endomorphisms [1] of this net, which form a C* tensor category $\mathscr{C}^{\mathrm{DHR}}(\mathscr{A})$ (strict, simple, with subobjects, direct sums and conjugates). The DHR endomorphisms localized in $O$, when restricted to $\mathscr{A}(O)$, are in fact endomorphisms of $\mathscr{A}(O)$, and they have the same intertwiners as endomorphisms of the net and as elements of $\mathrm{End}(\mathscr{A}(O))$ [2]. Therefore, they are the objects of a C* tensor category $\mathscr{C}^{\mathrm{DHR}}(\mathscr{A})|_O$, which is a full subcategory both of $\mathscr{C}^{\mathrm{DHR}}(\mathscr{A})$ and of $\mathrm{End}(N)$, $N = \mathscr{A}(O)$.

In other words, if $\rho$ is localized in $O$, then one may safely drop the distinction between $\rho \in \mathscr{C}^{\mathrm{DHR}}(\mathscr{A})$ and $\rho \in \mathscr{C}^{\mathrm{DHR}}(\mathscr{A})|_O \subset \mathrm{End}(N)$.

Since $\dim(\rho)$ was defined in terms of intertwiners, one may assign the same dimension to $\rho$ as a DHR endomorphism, and the same properties (additivity and multiplicativity) remain valid. This definition coincides [3] with the "statistical dimension" originally defined in terms of the statistics operators [1, 4].



M. Bischoff et al., *Tensor Categories and Endomorphisms of von Neumann Algebras*, SpringerBriefs in Mathematical Physics, DOI 10.1007/978-3-319-14301-9_3

It is physically most important that $\mathscr{C}^{\mathrm{DHR}}(\mathscr{A})$ is in fact a *braided* category, and in certain cases even *modular*. However, in our exposition, a braiding of the category $\mathscr{C}$ is not required before Sect. 4.5, and the braided category is not required to be modular before Sect. 4.11.

If the category $\mathscr{C} \subset \mathrm{End}_0(N)$ has only finitely many equivalence classes of irreducible objects (sectors), then it is called **rational**. In this case, the structures discussed below admit only finitely many realizations, with complete classification available in many models. The **global dimension** of a rational tensor category is

$$\dim(\mathscr{C}) := \sum_{[\rho]\,\mathrm{irr}} \dim(\rho)^2, \tag{3.0.1}$$

where the sum extends over the irreducible sectors of $\mathscr{C}$.

*Example 3.1* (The Ising tensor category) In order to illustrate the "rigidity" of a C* tensor category (and as a reference for further examples), we introduce the **Ising category**, which is one of two tensor categories with three self-conjugate equivalence classes [id], $[\tau]$, $[\sigma]$ of irreducible objects with "fusion rules" $[\tau^2] = [\mathrm{id}]$, $[\tau \circ \sigma] = [\sigma \circ \tau] = [\sigma]$, $[\sigma^2] = [\mathrm{id}] \oplus [\tau]$. It arises in QFT, e.g., as the category of DHR endomorphisms Chap. 5 of the chiral Ising model.

The tensor category is specified by a choice of a representative in each class, an isometric intertwiner in each intertwiner space according to the fusion rules, and the action of the representative endomorphisms on the intertwiners. For all unitarily equivalent endomorphisms, the intertwiners are canonically related. (To specify a category in this manner, is sometimes refereed to as the "Cuntz algebra approach".)

Because $\tau \circ \sigma$ is unitarily equivalent to $\sigma$, one can choose $\tau$ in its equivalence class such that $\tau \circ \sigma = \sigma$. Because $\tau^2$ is unitarily equivalent to the identity id and $\tau^2 \circ \sigma = \sigma$, it follows from irreducibility of $\sigma$ that $\tau^2 = \mathrm{id}$. Therefore, $\mathrm{Hom}(\sigma, \tau\sigma) = \mathrm{Hom}(\mathrm{id}, \tau^2) = \mathbb{C} \cdot \mathbf{1}$. The remaining nontrivial intertwiner spaces are spanned by a pair of orthogonal isometries $r \in \mathrm{Hom}(\mathrm{id}, \sigma^2)$ and $t \in \mathrm{Hom}(\tau, \sigma^2)$, satisfying $rr^* + tt^* = 1$, and $u \in \mathrm{Hom}(\sigma, \sigma\tau) = \mathrm{Hom}(\sigma, \sigma\tau) \times 1_\sigma = \mathrm{Hom}(\sigma^2, \sigma^2)$. Because $u^2 \in \mathrm{Hom}(\sigma, \sigma\tau^2) = \mathrm{Hom}(\sigma, \sigma) = \mathbb{C} \cdot \mathbf{1}$, one may choose $u = rr^* - tt^*$.

Because $\tau(r) \in \mathrm{Hom}(\tau, \sigma^2)$, one may choose $t = \tau(r)$, thus fixing the action of $\tau$:

$$\tau(r) = t, \quad \tau(t) = r, \quad \tau(u) = -u.$$

$\sigma(r) \in \mathrm{Hom}(\sigma, \sigma^3)$ and $\sigma(\tau) \in \mathrm{Hom}(\sigma\tau, \sigma^3)$ are linear combinations of $r$ and $t$, resp. $ru$ and $tu$, invariant under the action of $\tau$. Imposing $\sigma^2(a) = rar^* + t\,\tau(a)t^*$ for $a = r$ and $a = t$ suffices to fix all coefficients up to an overall sign. For the Ising category one has

$$\sigma(r) = 2^{-\frac{1}{2}}(r+t), \quad \sigma(t) = 2^{-\frac{1}{2}}(r-t)u.$$

(The category of DHR endomorphisms of the $su(2)$ current algebra at level 2 is specified by the opposite sign: $\sigma(r) = -2^{-\frac{1}{2}}(r+t), \sigma(t) = -2^{-\frac{1}{2}}(r-t)u$.)

The dimensions are $\dim(\tau) = 1$, $\dim(\sigma) = \sqrt{2}$, and the global dimension is $\dim(\mathscr{C}) = 4$.

## 3.1 C* Frobenius Algebras

A Frobenius algebra $\mathbf{A} = (\theta, w, x, \widehat{w}, \widehat{x})$ in a C* tensor category (satisfying the unit, counit, associativity, coassociativity and Frobenius relations [5]) is called **C* Frobenius algebra** if the dual morphisms are given by the adjoint operators: $\widehat{w} = w^*$ and $\widehat{x} = x^*$. By the latter property, the unit and counit relations become equivalent, and so do the associativity and coassociativity relations.

More precisely, $\theta$ is an object of the C* category, and $w \in \mathrm{Hom}(\mathrm{id}, \theta)$ and $x \in \mathrm{Hom}(\theta, \theta^2)$ are morphisms satisfying the relations

**unit property:** $(w^* \times 1_\theta) \circ x = (1_\theta \times w^*) \circ x = 1_\theta$

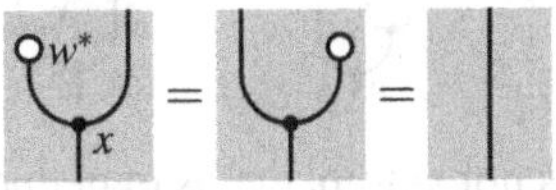


(3.1.1)

**associativity:** $(x \times 1_\theta) \circ x = (1_\theta \times x) \circ x$

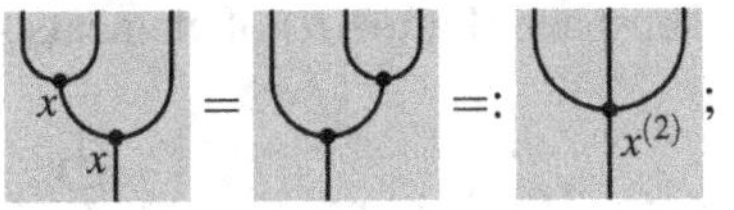


(3.1.2)

**Frobenius property:** $(1_\theta \times x^*) \circ (x \times 1_\theta) = xx^* = (x^* \times 1_\theta) \circ (1_\theta \times x)$

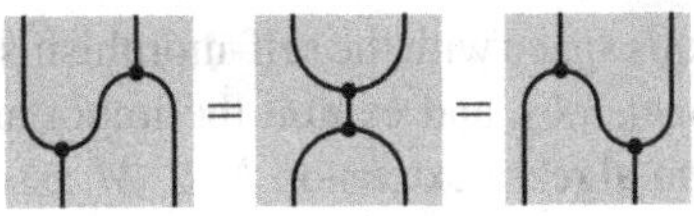


(3.1.3)

In view of Eq. (3.1.2), we also write $x^{(2)}$ for $(x \times 1_\theta) \circ x = (1_\theta \times x) \circ x$.

Clearly, $w^*w \in \mathrm{Hom}(\mathrm{id}, \mathrm{id}) = \mathbb{C}$ is a multiple of $\mathbf{1}$.

**Definition 3.2** If in addition, also $x^*x \in \mathrm{Hom}(\theta, \theta)$ is a multiple of $1_\theta$, the C* Frobenius algebra is called **special**. If moreover,

$$w^*w = d_{\mathbf{A}} \cdot 1_{\mathrm{id}} \quad \text{and} \quad x^*x = d_{\mathbf{A}} \cdot 1_\theta \tag{3.1.4}$$

with $d_{\mathbf{A}} = \sqrt{\dim(\theta)}$, we call the C* Frobenius algebra **standard**. The number $d_{\mathbf{A}} \geq 1$ is called the dimension of **A**.

If $\alpha : N \to M$ and $\overline{\alpha} : M \to N$ are conjugate homomorphisms between two factors, and $(w \in \mathrm{Hom}(\mathrm{id}_N, \overline{\alpha}\alpha), \overline{w} \in \mathrm{Hom}(\mathrm{id}_M, \alpha\overline{\alpha}))$ is a solution of the conjugacy relations, then

$$(\theta = \overline{\alpha}\alpha, w, x = \overline{\alpha}(\overline{w}))$$

is a C* Frobenius algebra. It is automatically special because $w^*w \in \mathrm{Hom}(\mathrm{id}_N, \mathrm{id}_N)$ and $\overline{w}^*\overline{w} \in \mathrm{Hom}(\mathrm{id}_M, \mathrm{id}_M)$ are multiples of **1**. $(\theta, w, x)$ is standard if and only if the pair $(w, \overline{w})$ is standard. Therefore, standardness can not always be enforced by a scalar rescaling of a special C* Frobenius algebra.

Our aim is to prove Theorem 3.11 which states that every standard C* Frobenius algebra is in fact of this type.

Let us first comment on the independence of the above axioms.

**Lemma 3.3** ([6]) *A Frobenius algebra is special, i.e.,* $x^*x = \lambda \cdot 1_\theta$, *if and only if* $x^*x \circ w = \lambda \cdot w$ *is a multiple of* $w$. *In particular, every Frobenius algebra with* Hom(id, $\theta$) *one-dimensional is special.*

*Proof* $x^*x = \lambda \cdot 1_\theta$ trivially implies $x^*x \circ w = \lambda \cdot w$. For the converse conclusion:

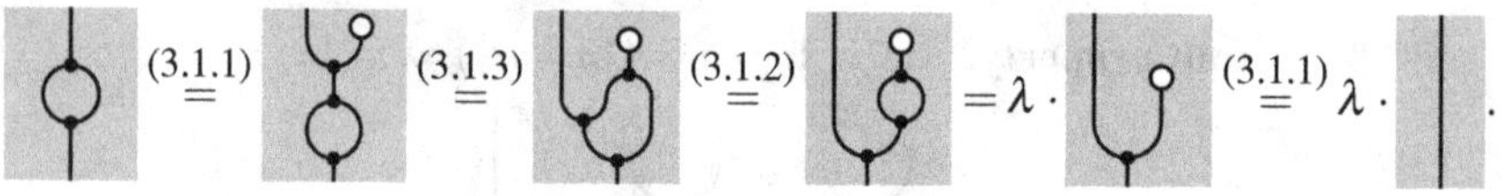


Standardness, however, is not automatic, as explained before.

**Definition 3.4** For a Frobenius algebra $(\theta, w, x)$ in a simple C* tensor category, $\mathrm{Hom}_0(\theta, \theta)$ is the subspace $\mathrm{Hom}(\theta, \theta)$ of elements satisfying

$$(1_\theta \times t) \circ x = x \circ t = (t \times 1_\theta) \circ x :$$

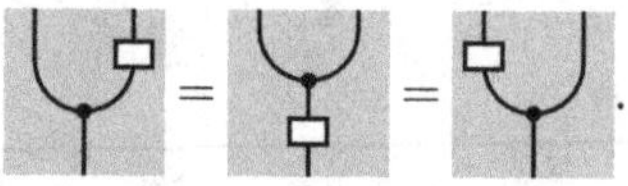

(3.1.5)

We shall later identify this space with the self-morphisms of the Frobenius algebra as a bimodule of itself (Sect. 3.6), and exhibit the importance of this space for the centre of the von Neumann algebra extension $N \subset M$ associated with a Frobenius algebra (Sect. 3.2).

**Lemma 3.5** *If* $(\theta, w, x)$ *is a Frobenius algebra in a simple C* tensor category, then* $n := x^*x$ *is a strictly positive element of* $\mathrm{Hom}_0(\theta, \theta)$.

*Proof* If $w^*w = d \cdot 1_{\mathrm{id}}$, then $d^{-1} \cdot ww^*$ is a projection, hence $1_\theta \geq d^{-1} \cdot ww^*$, hence

$$x^*x \geq d^{-1} \cdot x^* \circ (1 \times ww^*) \circ x \overset{(3.1.1)}{=} d^{-1} \cdot 1_\theta \tag{3.1.6}$$

is strictly positive. Equation (3.1.5) follows by associativity and the Frobenius property:

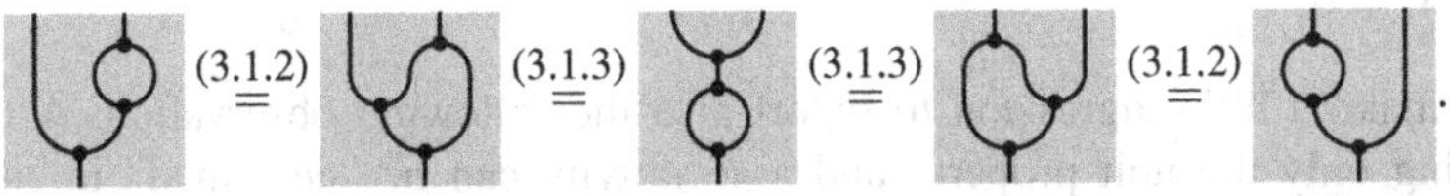


□

**Corollary 3.6** *The equivalent Frobenius algebra* $(\theta, \widehat{w}, \widehat{x})$ *with*

$$\widehat{w} := n^{\frac{1}{2}} \circ w, \quad \widehat{x} := (n^{-\frac{1}{2}} \times n^{-\frac{1}{2}}) \circ x \circ n^{\frac{1}{2}}$$

*is special.*

*Proof* Replacing $w, x$ by $\widehat{w}, \widehat{x}$ clearly preserves the unit property, associativity, and Frobenius property, and $\widehat{w}^*\widehat{w} \in \mathrm{Hom}(\mathrm{id}, \mathrm{id}) = \mathbb{C} \cdot 1$. Specialness follows because along with $n \in \mathrm{Hom}_0(\theta, \theta)$, also $n^{-1} \in \mathrm{Hom}_0(\theta, \theta)$, hence:

$$\widehat{x}^*\widehat{x} = n^{\frac{1}{2}} \circ x^* \circ (n^{-1} \times n^{-1}) \circ x \circ n^{\frac{1}{2}} \overset{(3.1.5)}{=} n^{-\frac{1}{2}} \circ x^* x \circ n^{-\frac{1}{2}}$$
$$= n^{-\frac{1}{2}} \circ n \circ n^{-\frac{1}{2}} = 1_\theta.$$

□

In C* tensor categories, also the Frobenius property is not independent from the other relations. We shall now prove that Eq. (3.1.3) follows from Eqs. (3.1.1) and (3.1.2) along with the special property $x^*x = \lambda \cdot 1$. Notice that specialness is a relation in $\mathrm{Hom}(\theta, \theta)$, and is thus "simpler" than the Frobenius relation Eq. (3.1.3) in $\mathrm{Hom}(\theta^2, \theta^2)$.

**Lemma 3.7** ([7]) *In a C* tensor category, the Frobenius property is a consequence of unit property, associativity, and specialness.*

*Proof* Let $X := (1_\theta \times x^*) \circ (x \times 1_\theta) - xx^* \in \mathrm{Hom}(\theta^2, \theta^2)$. Then, if $x^*x = d \cdot 1$, one has

$$X^*X = (x^* \times 1_\theta) \circ (1_\theta \times xx^*) \circ (x \times 1_\theta) - d \cdot xx^* \in \mathrm{Hom}(\theta^2, \theta^2),$$

where for the two mixed terms the associativity relation has been used. We define the map $\delta : \mathrm{Hom}(\theta^2, \theta^2) \to \mathrm{Hom}(\theta^2, \theta^2)$ given by

$$\delta(T) = (x^* \times 1_\theta) \circ (1_\theta \times T) \circ (x \times 1_\theta) :$$ 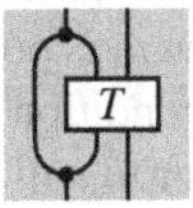


$\delta$ is positive and faithful: Namely if $T = Y^*Y$ is positive, then $\delta(T) = Z^*Z$ with $Z = (1_\theta \times Y) \circ (x \times 1_\theta)$, hence $\delta(Y^*Y)$ is positive; and $\delta(Y^*Y) = 0$ implies

$Z = 0$ from which it follows that $Y = 0$ by the unit property Eq. (3.1.1). We apply $\delta$ to $T = X^*X$. Again, using the associativity relation, one finds $\delta(X^*X) = 0$. Hence $X = 0$. □

We make a little digression to report also the following observation: A triple satisfying only the unit property and associativity can be "deformed" in such a way that it is in addition special. Then the Frobenius property also follows by Lemma 3.7, hence the deformed triple is a C* Frobenius algebra. There enters in the proof, however, a certain "regularity condition" which we do not quite know how to control.

The admitted deformations by any invertible element $n \in \mathrm{Hom}(\theta, \theta)$ are defined via

$$w \mapsto n^{*-1} \circ w, \quad x \mapsto (n \times n) \circ x \circ n^{-1},$$

obviously preserving Eqs. (3.1.1) and (3.1.2). The deformed triple is standard if

$$x^* \circ (n^*n \times n^*n) \circ x = n^*n.$$

We want to solve this eqution by iterating the following recursion:

$$m_{k+1} := x^* \circ (m_k \times m_k) \circ x,$$

starting with $m_0 = 1$, i.e., $m_1 = x^*x$. Clearly, each $m_k$ is a positive element of $\mathrm{Hom}(\theta, \theta)$. It is even strictly positive, because $(\theta, w_{k+1} := m_k^{-\frac{1}{2}} \circ w_k, x_{k+1} := (m_k^{\frac{1}{2}} \times m_k^{\frac{1}{2}}) \circ v_k \circ m_k^{-\frac{1}{2}})$ is a sequence of triples satisfying Eqs. (3.1.1) and (3.1.2), and $x_k^* x_k = m_k$ is strictly positive by Eq. (3.1.6). The question is, of course, whether $(m_k)_k$ converges.

Now $\mathrm{Hom}(\theta, \theta)$ equipped with the product $m_1 * m_2 = x^* \circ (m_1 \times m_2) \circ x$ is an algebra. The algebra has the unit $ww^*$, and is associative by Eq. (3.1.2). It is finite-dimensional, because $\mathrm{Hom}(\theta, \theta)$ is finite-dimensional. Hence it is isomorphic to some matrix algebra. W.r.t. this product, $m_0 = 1_\theta$, $m_1 = 1_\theta * 1_\theta$, and $m_k = 1_\theta^{*2^k}$. Because $m_k$ are strictly positive, they cannot be zero, hence $1_\theta$ is not nilpotent w.r.t. the $*$-product. Hence it has some largest eigenvalue, and hence some multiple $\mu_0$ of $1_\theta$ has a largest eigenvalue 1, so that $\mu_0^{*2^k}$ converges to an idempotent $m$ w.r.t. the $*$-product. This element therefore solves $x^* \circ (m \times m) \circ x = m$. If $m$ is strictly positive, then deforming the original triple $(\theta, w, x)$ with $n = m^{\frac{1}{2}}$, would give rise to a special triple, which then satisfies the Frobenius property by Lemma 3.7. However, we only know that $m$ is positive as a limit of strictly positive elements of $\mathrm{Hom}(\theta, \theta)$. The "regularity condition" mentioned above is the absence of a kernel of the limit. (Actually, in order to solve the equation, one may start from any initial element $\mu_0$ (not necessarily a multiple of $1_\theta$), but in the most general case, one will have even less control over the invertibility of the limit.)

After this digression, we return to the main line of the chapter.

## 3.2 Q-Systems and Extensions

**Definition 3.8** A **Q-system** is a standard Frobenius algebra $\mathbf{A} = (\theta, w, x)$ in a simple strict C* tensor category $\mathscr{C}$. Its *dimension* is $d_{\mathbf{A}} = \sqrt{\dim(\theta)}$.

Even in the irreducible case, where the canonical endomorphism $\theta$ fixes the intertwiner $w \in \mathrm{Hom}(\mathrm{id}, \theta)$ up to a complex phase, there may be finitely many inequivalent $x \in \mathrm{Hom}(\theta, \theta^2)$ [8].

From now on, we reserve the graphical representation

$w =$ , $x =$ 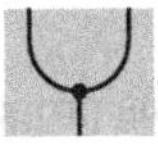, $r := x \circ w =$ 

for the intertwiners associated with a Q-system, i.e., $w$ and $x$ satisfy Eqs. (3.1.1)–(3.1.4), and $(r, r)$ satisfies Eq. (2.2.1). We shall freely use these properties in the sequel.

For the irreducible case, and $\mathscr{C} = \mathrm{End}_0(N)$, this definition first appeared in [9] as a characterization of subfactors $N \subset M$. In this section, we review and generalize this work to the reducible case. The correspondence between Q-systems and **extensions** of a factor (= inclusions into a (possibly non-factorial) von Neumann algebra) is the main reason for the study of Q-systems. In quantum field theory, Q-systems in $\mathscr{C} = \mathscr{C}^{\mathrm{DHR}}(\mathscr{A})$ correspond to extensions $\mathscr{A} \subset \mathscr{B}$ of a given QFT. Non-factorial extensions naturally arise, e.g., in the "universal construction" of boundary conditions discussed in [10], cf. Sect. 5.4.

An immediate consequence of standardness is the following:

**Corollary 3.9** *Let* $\mathbf{A} = (\theta, w, x)$ *be a Q-system,* $r = x \circ w$*. Then* $(r, \overline{r} = r)$ *is a standard pair for* $(\theta, \overline{\theta} = \theta)$*. The left and right Frobenius conjugations* $\mathrm{Hom}(\theta, \theta^2) \to \mathrm{Hom}(\theta^2, \theta)$, $y \mapsto (r^* \times 1_\theta) \circ (1_\theta \times y)$ *and* $y \mapsto (1_\theta \times r^*) \circ (y \times 1_\theta)$ *take* $x$ *to* $x^*$.

*Proof* The conjugacy relations Eq. (2.2.1) follow by applying the definition  $=$ 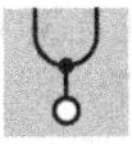 in several ways to Eqs. (3.1.3), and (3.1.1). $(r, \overline{r} = r)$ is a standard pair because $r^* r = w^* x^* x w = d_{\mathbf{A}} w^* w = d_{\mathbf{A}}^2 = \dim(\theta)$. □

*Remark 3.10* If $\mathbf{A} = (\theta, w, x)$ is only special, $w^* w = d_w \cdot 1$, $x^* x = d_x \cdot 1_\theta$, then $(r, r)$ still solves the conjugacy relations by the Frobenius and unit properties. Therefore, $r^* r \geq \dim(\theta)$ by the definition of the dimension as an infimum. Hence, $d_w d_x \geq \dim(\theta)$ with equality if and only if $\mathbf{A}$ is standard.

Let $N \subset M$ be an infinite subfactor of finite index, and $\iota : N \to M$ the embedding homomorphism. This gives rise to a Q-system in the C* tensor category $\mathrm{End}_0(N)$ as follows. Because the index $[M : N]$ is finite, the dimension $\dim(\iota)$ is finite, hence there is a conjugate homomorphism $\overline{\iota} : M \to N$. Let

$$w \in \mathrm{Hom}(\mathrm{id}_N, \overline{\iota}\iota) \subset N, \quad v \in \mathrm{Hom}(\mathrm{id}_M, \iota\overline{\iota}) \subset M$$

be a standard solution of the conjugacy relations Eq. (2.2.1). Then the triple

$$\mathbf{A} = (\theta, w, x), \quad \theta := \bar{\iota}\iota \in \mathrm{End}_0(N), \quad w \in N, \quad x := \bar{\iota}(v) \in N \tag{3.2.1}$$

is a Q-system in $\mathrm{End}_0(N)$ of dimension $d_{\mathbf{A}} = \dim(\iota)$. Graphically "resolving" $\theta = \bar{\iota} \circ \iota$, the intertwiners $w$ and $x = \bar{\iota}(v)$ are displayed as

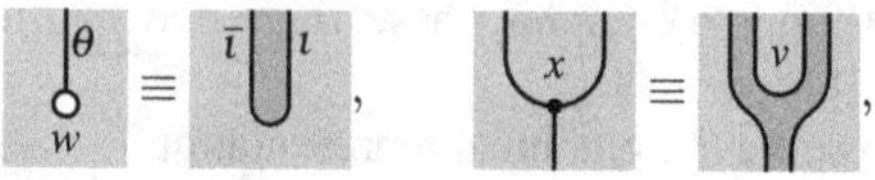


so that the unit, associativity and Frobenius properties are trivially satisfied:

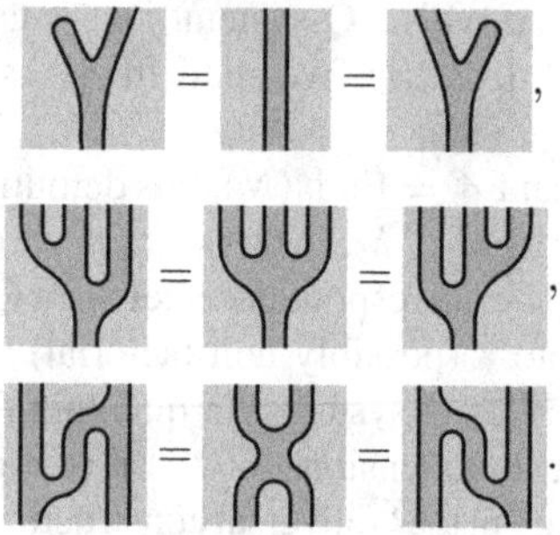

Notice that the projections $d_{\mathbf{A}}^{-1} \cdot ww^*$ and $d_{\mathbf{A}}^{-1} \cdot xx^*$ have the same properties as the Jones projections in the type II case [11], satisfying the Temperley-Lieb algebra and starting the "Jones tunnel". The Jones "planar algebra" [12] associated with a subfactor is the 2-category with two objects $N$ and $M$, whose 1-morphisms are sub-homomorphisms of alternating products of $\iota$ and $\bar{\iota}$, namely $\rho \prec (\bar{\iota}\iota)^n \in \mathrm{End}_0(N)$ for any $n \in \mathbb{N}$, $\varphi \prec \iota(\bar{\iota}\iota)^n \in \mathrm{Hom}(N, M)$, etc., and whose 2-morphisms are their intertwiners.

If $M = \bigoplus_i M_i$ is not a factor, and $\iota(n) = \bigoplus_i \iota_i(n)$ as in Sect. 2.3, then the Q-system defined by Eq. (3.2.1) can be computed with Proposition 2.7:

$$\theta(n) = \sum_i s_i \theta_i(n) s_i^*, \quad w = \sum_i \sqrt{\frac{d}{d_i}} \cdot s_i \circ w_i, \quad x = \sum_i \sqrt{\frac{d_i}{d}} \cdot (s_i \times s_i) \circ x_i \circ s_i^* \tag{3.2.2}$$

where $d = \sqrt{\dim(\theta)} = \sqrt{\sum_i \dim(\theta_i)}$, in compliance with Eq. (2.3.1).

The projections $p_i = s_i s_i^* \in \mathrm{Hom}(\theta, \theta)$ are elements of $\mathrm{Hom}_0(\theta, \theta)$, cf. Definition 3.4, i.e., they satisfy Eq. (3.1.5).

The central result of this section is the converse to the construction of a Q-system from an inclusion map $\iota : N \to M$: namely, the larger von Neumann algebra $M$ can be reconstructed from $N$ and the Q-system.

**Theorem 3.11** ([9]) *Let $N$ be an infinite factor, and $\mathbf{A} = (\theta, w, x)$ be a Q-system in* $\mathrm{End}_0(N)$. *Then there is a von Neumann algebra $M$ and a homomorphism $\iota : N \to M$ with conjugate $\bar{\iota} : M \to N$ such that $\theta = \bar{\iota}\iota$, and a standard solution $(w, v)$ of the conjugacy relations Eq. (2.2.1) such that $x = \bar{\iota}(v)$. The dimension $d_{\mathbf{A}}$ equals the dimension* $\dim(\iota) = \sqrt{\dim(\theta)}$.

*Proof* The algebra $M$ is reconstructed from $N$ and the Q-system by adjoining to $N$ one new element, called $v$, whose algebraic relations are the same as those of the operator $v \equiv v \times 1_\iota \in \mathrm{Hom}(\iota, \iota\bar{\iota}\iota) = \mathrm{Hom}(\iota, \theta\iota)$ if we *knew* that the Q-system comes from a conjugate solution $(w, v)$ as before. Namely, $v$ satisfies the commutation relations:

$$v\iota(n) = \iota\theta(n)v$$

with the elements $n \in N$ (where $\iota$ is the embedding of $N$ into the larger algebra $M$), i.e., $v \in \mathrm{Hom}(\iota, \iota\theta)$, its square is

$v^2 := \iota(x)v :$ 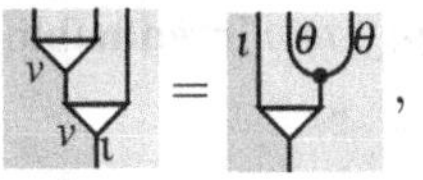 ,

and its adjoint is

$$v^* := \iota(w^* x^*)v.$$

It follows from these relations that every element of $M$ can be written in the form $\iota(n)v$ for some $n \in N$. The product thus defined is associative by virtue of Eq. (3.1.2), and it has a unit $\mathbf{1}_M = \iota(w^*)v$ by virtue of Eq. (3.1.1). The definition of $v^*$ implies the adjoint of a general element of $M$, namely $(\iota(n)v)^* = v^*\iota(n^*) = \iota(w^*x^*\theta(n^*))v$. This turns $M$ into a *-algebra, because the Frobenius property Eq. (3.1.3) ensures that the adjoint is an anti-multiplicative involution.

We have now constructed $M$ as a *-algebra. To see that it is in fact a von Neumann algebra, one has to induce the weak topology from $N$ to $M$ with the help of the faithful conditional expectation $\mu : M \to N$ given by

$$\mu(m) = d_{\mathbf{A}}^{-1} \cdot w^*\bar{\iota}(m)w, \quad \mu(vv^*) = d_{\mathbf{A}}^{-1} \cdot \mathbf{1}_N.$$

Here,

$$\bar{\iota}(\iota(n)v) := \theta(n)x$$

defines a conjugate homomorphism $\bar{\iota} : M \to N$, with $(w, v)$ as a standard solution of the conjugacy relations. $M$ is already weakly closed with respect to the induced topology because it is finitely generated from $N$. □

*Remark 3.12* It may be convenient to consider, rather than the single generator $v$ of the extension, the system of generators $\psi_\rho = \iota(w_\rho^*)v$ (charged intertwiners), where

$\rho \prec \theta$ is an irreducible sub-endomorphism, and $w_\rho \in \mathrm{Hom}(\rho, \theta)$. By definition, $\psi_\rho \in \mathrm{Hom}(\iota, \iota \circ \rho)$ which is equivalent to the commutation relations (suppressing the embedding map $\iota$)

$$\psi_\rho n = \rho(n)\psi_\rho \qquad (n \in N). \tag{3.2.3}$$

Every element of $M$ has an expansion $\sum n_\rho \psi_\rho$ into a basis of charged intertwiners with coefficients in $N$. The Q-system controls the product and adjoint of charged intertwiners.

In the sequel, we shall always use Q-systems to characterize extensions $N \subset M$ of a given factor $N$. In particular, all properties of the embedding are encoded in the Q-system, see also Chap. 4.

**Lemma 3.13** *For $\iota : N \to M$, the following are equivalent:*

(i) *The extension is irreducible: $\iota(N)' \cap M = \mathbb{C} \cdot \mathbf{1}_M$;*
(ii) *$\iota : N \to M$ is irreducible:* $\mathrm{Hom}(\iota, \iota) = \mathbb{C} \cdot \mathbf{1}_M$;
(iii) $\mathrm{dimHom}(\mathrm{id}_N, \bar{\iota}\iota) = 1$.

*Accordingly, we call a Q-system* **irreducible** *iff* $\mathrm{dimHom}(\mathrm{id}_N, \theta) = 1$.

*Example 3.14* (Q-systems of the Ising category) The Ising category (cf. Example 3.1) has two irreducible Q-systems: (id, 1, 1) with $M = N$, and $(\theta = \sigma^2, w = 2^{\frac{1}{4}} r, x = 2^{\frac{1}{4}} \sigma(r) = 2^{-\frac{1}{4}}(r + t))$. In the latter case, the extension is $M = \iota(N) \vee \psi$, where $\psi = 2^{\frac{1}{4}} \iota(t^*) v$ satisfies the relations $\psi\iota(n) = \iota(\tau(n))\psi$, $\psi^* = \psi$, $\psi^2 = 1$. $M$ has an automorphism (fixing $N$, = gauge transformation) $\alpha : \psi \mapsto -\psi$. The conjugate $\bar{\iota}$ in the latter case takes $\iota(n)$ to $\theta(n) = \sigma^2(n)$ and $\psi$ to $\sigma^2(t^*)(r + t) = rt^* + tr^*$.

For an irreducible Q-system, $M$ is automatically a factor, because $M' \cap M \subset \iota(N)' \cap M$. However, when $\mathrm{Hom}(\mathrm{id}_N, \theta)$ is more than one-dimensional, then $M$ may have a nontrivial centre, as characterized by (ii) of the following Lemma.

**Definition 3.15** We call the Q-system **simple**,[1] if the von Neumann algebra $M$ in Theorem 3.11 is a factor.

We shall see the equivalence of this definition with the usual one in Corollary 3.40.
In the sequel, we give various characterizations of the relative commutant $N' \cap M$ and of the centre of $M$.

**Lemma 3.16** (i) *The relative commutant $N' \cap M$ is given by the elements $\iota(q)v$, $q \in \mathrm{Hom}(\theta, \mathrm{id}_N)$.*

[1] The term **factorial** might be more appropriate in this context. "Simple", however, is more in line with standard category terminology, cf. Corollary 3.40.

(ii) *$\iota(q)v$ is idempotent iff* $(q\times q)\circ x=q$: 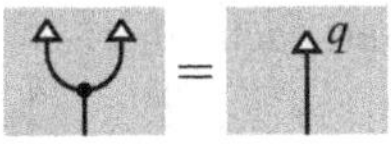 *, and it is selfadjoint iff* $q^*=(1_\theta\times q)\circ x\circ w$: 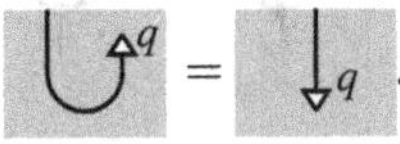.

(iii) *The centre of $M$ is given by the elements $\iota(q)v$, where $q$ belongs to the subspace of* $\mathrm{Hom}(\theta,\mathrm{id}_N)$ *of elements satisfying*

$$(q\times 1_\theta)\circ x=(1_\theta\times q)\circ x:$$ 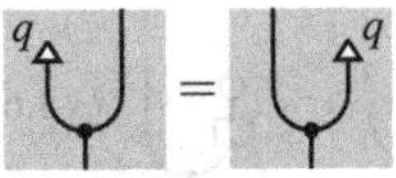 (3.2.4)

*In particular, the central projections are given by $\iota(q)v$ where $q\in\mathrm{Hom}(\theta,\mathrm{id}_N)$ satisfies all the relations in* (ii) *and* (iii).

*Proof* We use the uniqueness of the representation $m=\iota(n)v$ for all three statements. Thus we write $c=\iota(q)v$ and characterize the properties of $c$ in terms of $q$:

(i) For $c\in\iota(N)'\cap M$, the commutation relation $c\iota(n)=\iota(n)c$ reads $\iota(q\theta(n))v=\iota(nq)v$. This is equivalent to $q\theta(n)=nq$.
(ii) Immediate from $(\iota(q)v)^2=\iota(q\theta(q)x)v$ and $(\iota(q)v)^*=\iota(w^*x^*\theta(q^*))v$.
(iii) The commutation relation $cv=vc$ for $c\in M'\cap M$ reads $\iota(qx)v=\iota(\theta(q)x)v$, hence $qx=\theta(q)x$. □

**Lemma 3.17** (i) *The linear maps* $\mathrm{Hom}(\theta,\mathrm{id}_N)\to\mathrm{Hom}(\theta,\theta)$, 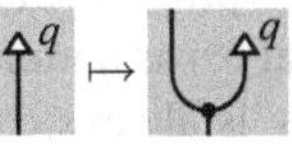, *and* $\mathrm{Hom}(\theta,\theta)\to\mathrm{Hom}(\theta,\mathrm{id}_N)$, 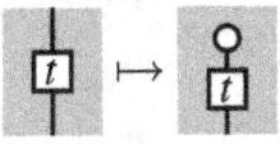 *define a bijection between* $\mathrm{Hom}(\theta,\mathrm{id}_N)$ *and the subspace of* $\mathrm{Hom}(\theta,\theta)$ *of elements satisfying the first of Eq.* (3.1.5)*:*

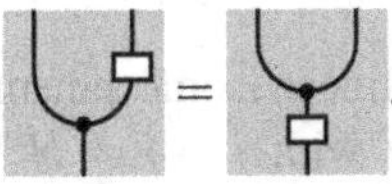 (3.2.5)

(ii) *$q\in\mathrm{Hom}(\theta,\mathrm{id}_N)$ satisfies Eq.* (3.2.4) *iff $t\in\mathrm{Hom}(\theta,\theta)$ (its image under the bijection in* (i)) *satisfies also*

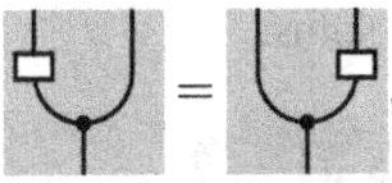 (3.2.6)

*i.e., iff $t\in\mathrm{Hom}_0(\theta,\theta)$.*

*Proof* (i) $\theta(q)x$ satisfies Eq. (3.2.5): By associativity 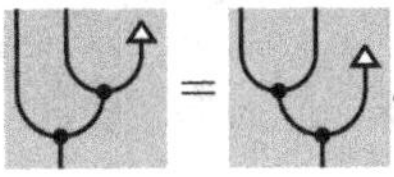.

The two maps invert each other:

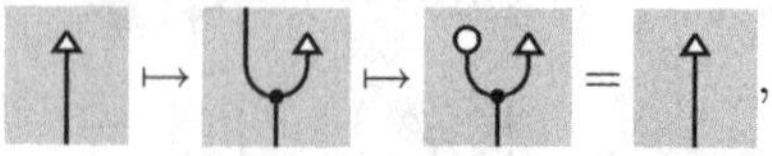

and

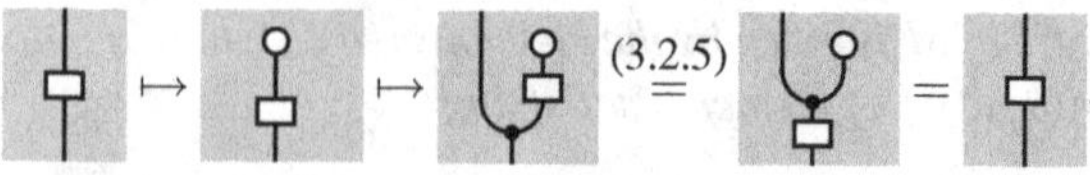


(ii) "If": 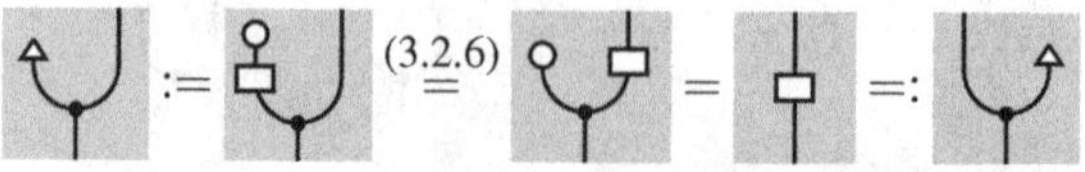

by the unit property.

"Only if": 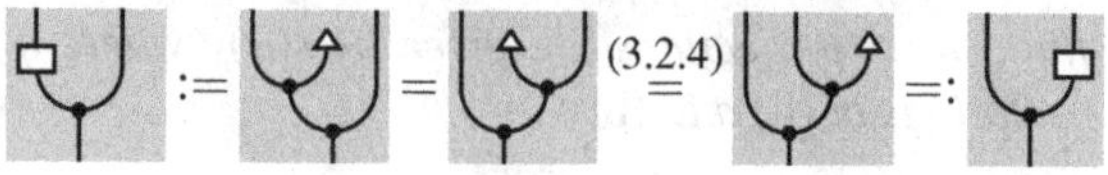

by associativity. □

Thus, the relative commutant $N' \cap M$ and the centre $M' \cap M$ are equivalently characterized by certain elements of $\mathrm{Hom}(\theta, \mathrm{id}_N)$ or of $\mathrm{Hom}(\theta, \theta)$. In particular, the space $\mathrm{Hom}_0(\theta, \theta)$, Definition 3.4, is one way to characterize the centre of $M$. We shall come back to this in Sects. 4.2 and 4.3.

*Remark 3.18* The standardness property of the Q-system is not used in the construction of the algebra $M$ in the proof of Theorem 3.11, and neither the (weaker) specialness property that $x^*x$ is a multiple of $1_\theta$. These properties are only required to ensure that $v^*v$ is a multiple of $\mathbf{1}_M$, namely $x^*x = \bar{\iota}(v^*v)$. Because $M' \cap M = \mathrm{Hom}(\iota\bar{\iota}, \iota\bar{\iota})$, $v^*v$ is always central in $M$, hence specialness is automatically satisfied if $M$ is a factor.

## 3.3 The Canonical Q-System

Let $j : N \to j(N)$ be an antilinear isomorphism of factors. E.g., $j : n \mapsto n^*$ is an antilinear isomorphism of $N$ with $j(N) = N^{\mathrm{opp}}$ (the algebra with the opposite product), or a Tomita conjugation $j = \mathrm{Ad}_J$ is an antilinear isomorphism of $N$ with $j(N) = N'$. For $\mathscr{C} \subset \mathrm{End}_0(N)$, let $j(\mathscr{C})$ the category with objects $\rho^j \equiv j \circ \rho \circ j^{-1}$ $(\rho \in \mathscr{C})$ and with intertwiners $j(t)$.

We denote by $\mathscr{C}_1 \boxtimes \mathscr{C}_2$ (the Deligne product) the completion of the tensor product of categories $\mathscr{C}_1 \otimes \mathscr{C}_2$ by direct sums.

**Proposition 3.19** ([13]) *If $\mathscr{C}$ has only finitely many inequivalent irreducible objects $\rho$, then there is a canonical irreducible Q-system* $\mathbf{R}$ *in $\mathscr{C} \boxtimes j(\mathscr{C})$ with*

$$[\Theta_{\mathrm{can}}] = \bigoplus_\rho [\rho] \otimes [\rho^j],$$

*where the sum runs over the equivalence classes of irreducible objects of $\mathscr{C}$. Its dimension is given by $d_{\mathbf{R}}^2 = \dim(\Theta_{\mathrm{can}}) = \dim(\mathscr{C})$. Choosing isometries $T_\rho \in \mathrm{Hom}(\rho \otimes \rho^j, \Theta_{\mathrm{can}})$, the Q-system is given by*

$$W = d_{\mathbf{R}}^{\frac{1}{2}} \cdot T_{\mathrm{id}}, \qquad X = d_{\mathbf{R}}^{-\frac{1}{2}} \sum_{\rho,\sigma,\tau} \left(\frac{d_\rho d_\sigma}{d_\tau}\right)^{\frac{1}{2}} \cdot (T_\rho \times T_\sigma) \circ \Big(\sum_a t_a \otimes j(t_a)\Big) \circ T_\tau^*,$$

*where the first sum extends over representatives of all sectors, and the inner sums over $a$ extend over orthonormal bases of isometries $t_a \in \mathrm{Hom}(\tau, \rho\sigma)$.*

Because of the anti-linearity of $j$, the sums over $a$ do not depend on the choice of orthonormal bases $t_a$. A different choice of $T_\rho$ gives a unitarily equivalent Q-system.

One easily proves (cf. Proposition 2.6)

**Lemma 3.20** ([7]) *Choosing, for every $\rho \in \mathscr{C}$, a conjugate $\overline{\rho} \in \mathscr{C}$ and a standard pair $(w, \overline{w})$, the assignment*

$$\rho \mapsto \overline{\rho}, \quad t \mapsto j\Big(\ \ \Big) = j\Big(\ \ \Big)$$

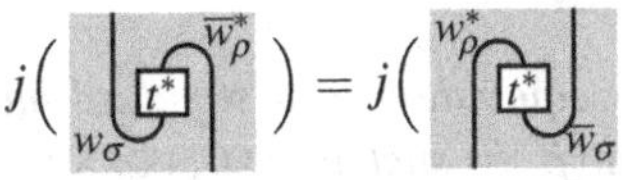


*taking* $\mathrm{Hom}(\rho, \sigma)$ *into* $\mathrm{Hom}(\overline{\rho}^j, \overline{\sigma}^j)$ *is a linear isomorphism between the $C^*$ tensor categories $\mathscr{C}^{\mathrm{opp}}$ and $j(\mathscr{C})$ (the category equipped with the opposite monoidal product).*

**Corollary 3.21** *The opposite tensor category $\mathscr{C}^{\mathrm{opp}}$ can be realized as $j(\mathscr{C}) \subset \mathrm{End}_0(N^{\mathrm{opp}})$ or $\mathrm{End}_0(N')$. Under this isomorphism, the canonical Q-system in $\mathscr{C} \boxtimes j(\mathscr{C})$ becomes a Q-system in $\mathscr{C} \boxtimes \mathscr{C}^{\mathrm{opp}}$ with $[\Theta] = \bigoplus [\rho] \otimes [\overline{\rho}]$.*

This is the way it is defined in the abstract setting (e.g., [15, Prop. 4.1]).

## 3.4 Modules of Q-Systems

A **module** ($\equiv$ left module) of a Q-system $\mathbf{A} = (\theta, w, x)$ is a pair $\mathbf{m} = (\beta, m)$, where $\beta$ is an object of the underlying category and $m \in \mathrm{Hom}(\beta, \theta\beta)$,[2] satisfying the relations

**unit property:** $(w^* \times 1_\beta) \circ m = 1_\beta$

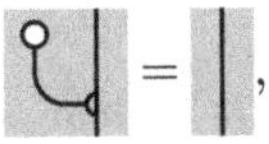

(3.4.1)

[2] More precisely, $(\beta, m^*)$ is a module and $(\beta, m)$ is a co-module. We do not make the distinction because the dualization is canonically given by the operator adjoint.

**representation property:** $(1_\theta \times m) \circ m = (x \times 1_\beta) \circ m$

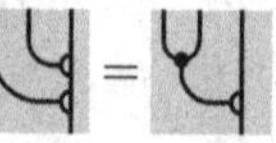

(3.4.2)

A module of a Q-system is called a **standard module** if $m^*m$ is a multiple of $1_\beta$. (This property is automatic if $(\beta, m)$ is irreducible as a module, and in particular if $\beta$ is irreducible as an endomorphism.)

A Q-system **A** is also a standard **A**-module ($\beta = \theta, m = x$).

Two modules $(\beta, m)$ and $(\beta', m')$ are **equivalent**, when there is an invertible $n \in \mathrm{Hom}(\beta, \beta')$ such that $m' \circ n = (1_\theta \times n) \circ m$. They are unitarily equivalent if there is a unitary such $n$.

**Lemma 3.22** (i) *If a module* $\mathbf{m} = (\beta, m)$ *is standard, then (with* $d_\mathbf{A} =$ *the dimension of the Q-system)*

$$m^*m =$$ 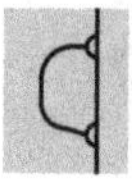 $$= d_\mathbf{A} \cdot 1_\beta.$$

(3.4.3)

(ii) *Every module is equivalent to a standard module, i.e., there is an invertible element* $n$ *of* $\mathrm{Hom}(\beta, \beta)$ *such that* $(\beta, (1_\theta \times n) \circ m \circ n^{-1})$ *is a standard module.*

*Proof* (i) follows from the representation property Eq. (3.4.2) and $x^*x = d_\mathbf{A} \cdot 1_\theta$, which imply

$$m^* \circ (1_\theta \times m^*m) \circ m =$$ 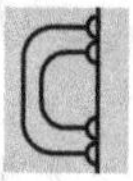 $=$ 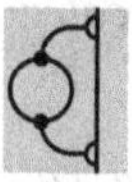 $= d_\mathrm{A} \cdot$  $= d_\mathrm{A} \cdot m^*m.$

For (ii), first we notice that $m^*m$ is an invertible positive element of $\mathrm{Hom}(\beta, \beta)$, because $e = d_\mathbf{A}^{-1} \cdot ww^*$ is a projection in $\mathrm{Hom}(\theta, \theta)$, hence by the unit property,

$$m^*m \geq m^* \circ (e \times 1_\beta) \circ m = d_\mathbf{A}^{-1} \cdot 1_\beta.$$

Let $n \in \mathrm{Hom}(\beta, \beta)$ be the square root of $m^*m$. Then by the representation property,

$$n^{-1}m^* \circ (1_\theta \times n^2) \circ mn^{-1} = n^{-1}m^* \circ (x^*x \times 1_\beta) \circ mn^{-1} = d_\mathbf{A} \cdot 1_\beta. \qquad \square$$

**Lemma 3.23** *If* $(\beta, m)$ *is a standard module, then in addition to the unit and representation relations, the relation*

$$(x^* \times 1_\beta) \circ (1_\theta \times m) = mm^* = (1_\theta \times m^*) \circ (x^* \times 1_\beta):$$

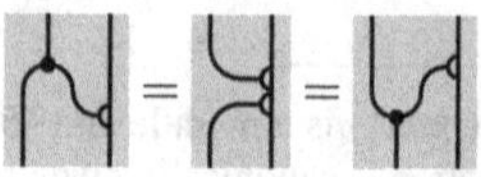

(3.4.4)

*holds. This implies*

$$m^* = (r^* \times 1_\beta) \circ (1_\theta \times m) :$$ 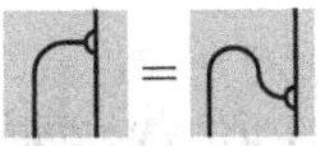 , (3.4.5)

*and consequently*

$$E := d_{\mathbf{A}}^{-1} \cdot (x^* \times 1_\beta) \circ (1_\theta \times m) = d_{\mathbf{A}}^{-1} \cdot$$ 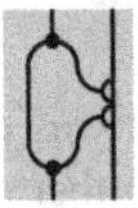

*is a self-adjoint idempotent, i.e., a projection in* $\mathrm{Hom}(\theta\beta, \theta\beta)$.

*Proof* The proof is very much the same as the proof of the Frobenius property in Lemma 3.7, with $X$ replaced by $X' := (1_\theta \times m^*) \circ (x \times 1_\beta) - mm^* \in \mathrm{Hom}(\theta\beta, \theta\beta)$, the associativity property of $x$ replaced by the representation property of $m$, and $\delta$ replaced by the faithful positive map $\delta' : \mathrm{Hom}(\theta\beta, \theta\beta) \to \mathrm{Hom}(\theta\beta, \theta\beta)$

$$\delta'(T) = (x^* \times 1_\beta) \circ (1_\theta \times T) \circ (x \times 1_\beta) :$$ 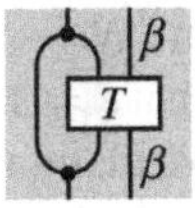


The equation for $m^*$ then follows by left composition with $w^* \times 1_\beta$, and the statement about $E$ follows because $E = d_{\mathbf{A}}^{-1} \cdot mm^*$ and $m^* m = d_{\mathbf{A}} \cdot 1_\beta$. □

From now on, we reserve the graphical representation

$$m =$$ 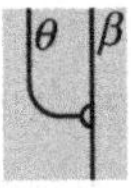


for the intertwiner associated with a standard module $\mathbf{m} = (\beta, m)$, i.e., $m$ satisfies Eqs. (3.4.1)–(3.4.3), hence also Eq. (3.4.5). We shall freely use these properties in the sequel.

If $\mathbf{A}$ is a Q-system in $\mathscr{C} = \mathrm{End}_0(N)$, corresponding to an extension $\iota : N \to M$, then every homomorphism $\varphi : N \to M$ of finite dimension gives rise to a standard module

$$(\beta, m) \equiv (\bar{\iota}\varphi, 1_{\bar{\iota}} \times v \times 1_\varphi) :$$ 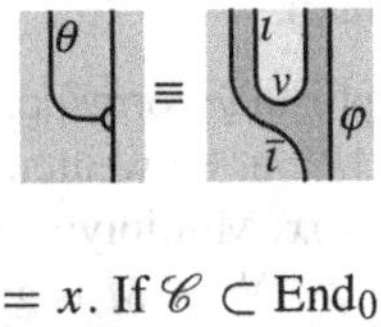 (3.4.6)


of $\mathbf{A}$. Notice that, as an operator in $N$, $m = \bar{\iota}(v) = x$. If $\mathscr{C} \subset \mathrm{End}_0(N)$ as specified in the beginning of the chapter, then the same is true provided $\bar{\iota}\varphi$ belongs to $\mathscr{C}$. This restriction on $\varphi$ is equivalent to the condition that $\varphi \prec \iota\rho$ with some $\rho \in \mathscr{C}$.

The converse is also true: namely, we prove now that every standard module is of this form:

**Proposition 3.24** *Every standard module* $\mathbf{m} = (\beta, m)$ *of a simple Q-system* $\mathbf{A} = (\theta, w, x)$ *in* $\mathrm{End}_0(N)$ *is unitarily equivalent to a standard module of the form* $(\bar{\iota}\varphi, x)$ *as in Eq. (3.4.6), where* $\varphi$ *is a homomorphism* $\varphi : N \to M$.

(The same result was derived by [16, Lemma 3.1] by an exhaustion argument, using the known number of modules in the case of a *braided* category; our proof is more constructive, and does not refer to a braiding.)

*Proof* Writing as before $\theta = \bar{\iota}\iota$, $m$ defines by left Frobenius conjugation an element

$$e = d_{\mathbf{A}}^{-1} \cdot (v^* \times 1_{\iota\beta}) \circ (1_\iota \times m)$$

of $\mathrm{Hom}(\iota\beta, \iota\beta) \subset M$. Then $1_{\bar{\iota}} \times e$ equals, by Eq. (3.4.4), the projection $E = d_{\mathbf{A}}^{-1} \cdot mm^*$ in Lemma 3.23, hence $e$ is also a projection. Let $\varphi \prec \iota\beta$ be the sub-homomorphism : $N \to M$ corresponding to this projection, and $s \in \mathrm{Hom}(\varphi, \iota\beta)$ an isometry such that $e = ss^*$. By left Frobenius conjugation, $\tilde{s} := (1_{\bar{\iota}} \times s^*) \circ (w \times 1_\beta) \in \mathrm{Hom}(\beta, \bar{\iota}\varphi)$. We claim that the range projection of $\tilde{s}$ equals $1_{\bar{\iota}\varphi}$.

Indeed, by inverting the definition of $e$, we have that

$$m = d_{\mathbf{A}} \cdot (1_{\bar{\iota}} \times ss^*) \circ (w \times 1_\beta),$$

hence

$$\tilde{s} = d_{\mathbf{A}}^{-1} \cdot (1_{\bar{\iota}} \times s^*) \circ m.$$

Now, we use again Eq. (3.4.4): $mm^* = d_{\mathbf{A}} \cdot 1_{\bar{\iota}} \times e = d_{\mathbf{A}} \cdot 1_{\bar{\iota}} \times ss^*$ to conclude

$$\tilde{s}\tilde{s}^* = d_{\mathbf{A}}^{-2} \cdot (1_{\bar{\iota}} \times s^*) \circ mm^* \circ (1_{\bar{\iota}} \times s^*) = d_{\mathbf{A}}^{-1} \cdot (1_{\bar{\iota}} \times s^*ss^*s) = d_{\mathbf{A}}^{-1} \cdot 1_{\bar{\iota}\varphi}.$$

Thus, while $\varphi \prec \iota\beta$ by construction, we also have $\iota\beta \prec \varphi$, hence $\beta$ is equivalent to $\bar{\iota}\varphi$. It follows that $u := \sqrt{d_{\mathbf{A}}} \cdot \tilde{s}$ is a unitary $u \in \mathrm{Hom}(\beta, \bar{\iota}\varphi)$. Then, inserting $s = (1_\iota \times \tilde{s}^*) \circ (w \times 1_\varphi)$ into $m = d_{\mathbf{A}} \cdot (1_{\bar{\iota}} \times ss^*) \circ (w \times 1_\beta)$, one arrives at

$$m = (1_\theta \times u^*) \circ (1_{\bar{\iota}} \times v \times 1_\varphi) \circ u.$$

This proves the claim. □

The homomorphism $\varphi$ corresponding to a module $\mathbf{m}$ can be explicitly computed: namely, $\varphi(n) \in M$ can be written as $\varphi(n) = \iota(k)v$ for some $k \in N$. Applying $\bar{\iota}$ implies $\beta(n) = \theta(k)x$. Multiplying $w^*$ from the right, implies $w^*\beta(n) = w^*\theta(k)x = kw^*x = k$. Hence $\varphi : N \to M$ is given by

$$\varphi(n) = \iota(w^*\beta(n))v \in M.$$

Considering **A** as a standard **A**-module ($\beta = \theta, m = x$), the corresponding homomorphism is $\varphi = \iota : N \to M$.

The modules of a Q-system $(\theta, w, x)$ are the objects of the **module category**. A **morphism** between two modules $(\beta, m)$ and $(\beta', m')$ is an element $t \in \mathrm{Hom}(\beta, \beta')$ satisfying

$$(1_\theta \times t) \circ m = m' \circ t : \tag{3.4.7}$$

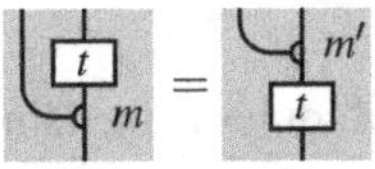


It is obvious from the definition that the modules are closed under right tensoring with $\rho \in \mathscr{C}$, namely $\mathbf{m} \times 1_\rho \equiv (\beta \circ \rho, \mathbf{m} \times 1_\rho)$ is again a module, and the corresponding homomorphism is $\varphi \circ \rho$. Moreover, the right tensoring is compatible with the morphisms. The category thus defined is therefore a right module category in the sense of [17, Definition 6].

Clearly, every $s \in \mathrm{Hom}(\varphi, \varphi')$ defines a morphism $t = 1_{\bar\iota} \times s$ between the associated standard modules. The converse is also true:

**Proposition 3.25** *Every morphism t between two standard modules* $(\bar\iota\varphi, m = 1_{\bar\iota} \times v \times 1_\varphi)$ *and* $(\bar\iota\varphi', m' = 1_{\bar\iota} \times v \times 1_{\varphi'})$ *is of the form* $t = 1_{\bar\iota} \times s$ *where* $s \in \mathrm{Hom}(\varphi, \varphi')$.

*Proof* $s = d_{\mathbf{A}}^{-1} \cdot \mathrm{LTr}_{\bar\iota}(t)$ does the job:

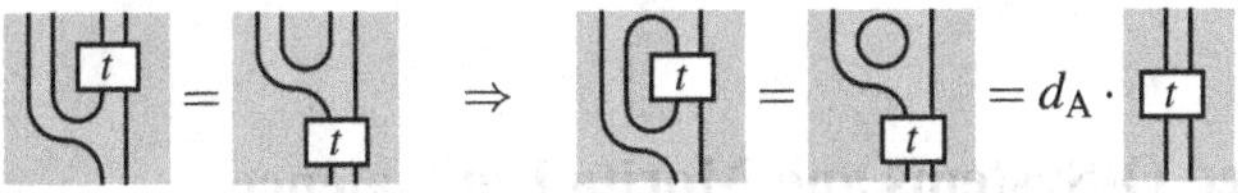


□

We recognize that the argument in the proof of Lemma 3.17 is just an instance of this general fact, namely Eq. (3.2.5) just states that $t \in \mathrm{Hom}(\theta, \theta)$ is a morphism between $\mathbf{A} = (\theta, x)$ as a **A**-module and itself, hence $t = 1_{\bar\iota} \times s = \bar\iota(s)$ with $s = \iota(q)v \in \mathrm{Hom}(\iota, \iota)$.

**Corollary 3.26** *The module category of a simple Q-system in* $\mathscr{C} \subset \mathrm{End}_0(N)$ *is equivalent to the full subcategory of* $\mathrm{Hom}(N, M)$ *whose objects are the homomorphisms* $\varphi \prec \iota\rho : N \to M$, $\rho \in \mathscr{C}$.

In particular:

**Corollary 3.27** *Let* $\mathbf{m} = (\beta, m)$ *be a reducible module. The space of self-morphisms of* $\mathbf{m}$ *is a finite-dimensional* $C^*$ *algebra. If* $p_i$ *are minimal projections in this algebra,* *and* $p_i = t_i t_i^*$ *with isometries* $t_i$, *then* $\mathbf{m} \simeq \bigoplus_i \mathbf{m}_i$ *with* $\mathbf{m}_i = (\beta_i, m_i)$, *where* $\beta_i = t_i^* \beta(\cdot) t_i$ *and* $m_i = (1_\theta \times t_i^*) \circ m \circ t_i$, *i.e.*,

$$\beta = \sum_i t_i \beta_i(\cdot) t_i^*, \quad m = \sum_i (1_\theta \times t_i) \circ m_i \circ t_i^*.$$

*Example 3.28* (Modules in the Ising category) The irreducible modules of the trivial Q-system are $(\rho, 1)$ with $\rho = \mathrm{id}, \sigma, \tau$. The corresponding homomorphisms : $N \to M = N$ are $\varphi = \rho$.

The modules $(\beta, x = 2^{-\frac{1}{4}}(r + t))$ of the nontrivial Q-system given in Example 3.14 ($\theta = \sigma^2 \simeq \mathrm{id} \oplus \tau$) arising from $\varphi \prec \iota\rho$ are:

(i) $\rho = \mathrm{id}$: module $(\sigma^2, x)$, homomorphism $\varphi = \iota$.
(ii) $\rho = \tau$: module $(\sigma^2 \tau, x)$, homomorphism $\varphi = \iota \circ \tau$.
(iii) $\rho = \sigma$: The module $(\beta = \theta\sigma = \sigma^3, x)$ is reducible: $\simeq (\beta_1 = \sigma, x) \oplus (\beta_2 = \sigma\tau, x)$ (with morphisms $\sigma(r) \in \mathrm{Hom}(\beta_1, \beta)$ and $\sigma(t) \in \mathrm{Hom}(\beta_2, \beta)$, respectively). For the submodule $(\sigma, x)$, one computes $\varphi_1 : n \mapsto r^* \sigma(n)(r + t\psi)$, in particular, $r \mapsto 2^{-\frac{1}{2}}(r + t\psi)$, $t \mapsto 2^{-\frac{1}{2}}(r - t\psi)$, $u \mapsto \psi$. For the submodule $(\sigma\tau, x)$, $\varphi_2 : n \mapsto r^* \sigma\tau(n)(r + t\psi)$, in particular $r \mapsto 2^{-\frac{1}{2}}(r - t\psi)$, $t \mapsto 2^{-\frac{1}{2}}(r + t\psi)$, $u \mapsto -\psi$. These homomorphisms are surjective, hence isomorphisms, and $\varphi_2 = \varphi_1 \circ \tau = \alpha \circ \varphi_1$ ($\alpha$ = gauge transformation $\psi \to -\psi$).

## 3.5 Induced Q-Systems and Morita Equivalence

Let $\mathbf{A} = (\theta, w, x)$ be a Q-system, defining an extension $\iota : N \to M$.

If $\mathbf{m} = (\beta, m)$ is a standard module of $\mathbf{A}$, and $\varphi : N \to M$ the corresponding homomorphism, we choose a conjugate homomorphism $\overline{\varphi} : M \to N$ and a solution of the conjugacy relations $w_\varphi$, $v_\varphi$. Then

$$\mathbf{A}_\varphi = (\theta_\varphi, w_\varphi, x_\varphi) \quad \text{with} \quad \theta_\varphi = \overline{\varphi}\varphi, \quad x_\varphi = \overline{\varphi}(v_\varphi)$$

is a Q-system. We call $\mathbf{A}_\varphi$ the **Q-system induced by m.**

Notice that, by definition, $\overline{\varphi}\varphi = \theta_\varphi = \overline{\iota}_\varphi \iota_\varphi$; but the corresponding extension $\iota_\varphi : N \to M_\varphi$ should not be confused with the homomorphism $\varphi : N \to M$, because $\iota_\varphi(n) = n \in N \subset M_\varphi$, while $\varphi(n) \neq n \in N \subset M$.

**Lemma 3.29** *If a Q-system* $\mathbf{A}_2 = (\theta_2, w_2, x_2)$ *is induced by a standard module* $(\beta_1, m_1)$ *of* $\mathbf{A}_1 = (\theta_1, w_1, x_1)$, *then* $\mathbf{A}_1$ *is induced by a standard module of* $\mathbf{A}_2$.

*Proof* By Proposition 3.24, $(\beta_1, m_1)$ is of the form $\beta_1 = \bar{\iota}_1\varphi_1$ and $m_1 = x_1$, where $\varphi_1 \prec \iota_1\rho$ for some $\rho \in \mathscr{C}$. By definition of the induced Q-system, $\iota_2 = \varphi_1$, and hence $\iota_1 \prec \iota_2\bar{\rho}$. Therefore, $\mathbf{A}_1$ is induced by the module $(\beta_2 = \bar{\iota}_2\varphi_2, m_2 = x_2)$ of $\mathbf{A}_2$, where $\varphi_2 = \iota_1$. □

**Definition 3.30** ([17, Definition 10]) Two Q-systems in $\mathscr{C}$ are **Morita equivalent** if their module categories are equivalent, i.e., there exists an invertible functor between the two module categories that commutes with the right tensoring by $\rho \in \mathscr{C}$.

**Proposition 3.31** *Two Q-systems are Morita equivalent if and only if one of them is induced by a standard module of the other one (which implies also the converse).*

*Proof* If $\mathbf{A}_2 = (\theta_2, w_2, x_2)$ is induced by a standard module $(\beta_1, m_1)$ of $\mathbf{A}_1 = (\theta_1, w_1, x_1)$, then $\iota_2 = \varphi_1 \prec \iota_1\rho$ for some $\rho \in \mathscr{C}$. Then the sub-homomorphisms $\varphi$ of $\iota_2\rho'$ for some $\rho' \in \mathscr{C}$ are the same as the sub-homomorphisms of $\iota_1\rho''$ for some $\rho'' \in \mathscr{C}$. Then, by Propositions 3.24 and 3.25, mapping the standard modules $(\bar{\iota}_1\varphi, x_1)$ of $\mathbf{A}_1$ to $(\bar{\iota}_2\varphi, x_2)$, and morphisms $t_1 = 1_{\bar{\iota}_1} \times s$ to $t_2 = 1_{\bar{\iota}_2} \times s$, defines a bijective functor that commutes with the right tensoring by $\rho \in \mathscr{C}$.

Conversely, if $\mathbf{A}_1$ and $\mathbf{A}_2$ are Morita equivalent, then there is a module $\mathbf{m}_1$ of $\mathbf{A}_1$ mapped by the bijective functor $F$ to $\mathbf{A}_2$ as a module of itself. By Proposition 3.24, $\mathbf{m}_1 = (\bar{\iota}_1\varphi, x_1)$ (up to equivalence) with $\varphi \prec \iota\rho$ for some $\rho \in \mathscr{C}$. We have to show that the Q-system $\mathbf{A}_\varphi$ induced by $\varphi$ is equivalent to $\mathbf{A}_2$. We first show that $\theta_\varphi = \bar{\iota}_\varphi\iota_\varphi = \bar{\varphi}\varphi$ equals $\theta_2$ (up to unitary equivalence).

For every $\sigma \in \mathscr{C}$, one has $\mathrm{Hom}(\sigma, \bar{\varphi}\varphi) \sim \mathrm{Hom}(\varphi\,\sigma, \varphi)$ by Frobenius reciprocity. Because $\varphi$ corresponds to $\iota_2$ under $F$, and $F$ commutes with right tensoring by $\sigma \in \mathscr{C}$, we further have $\mathrm{Hom}(\varphi\,\sigma, \varphi) \sim \mathrm{Hom}(\iota_2\,\sigma, \iota_2) \sim \mathrm{Hom}(\sigma, \theta_2)$, from which the claim follows.

Since the construction of the induced Q-system is invariant under the isomorphism of module categories $F$, it follows that the Q-system induced by $\varphi$ from $\mathbf{A}_1$ coincides with the Q-system induced by $\iota_2$ from $\mathbf{A}_2$, which is of course $\mathbf{A}_2$. □

Thus, the Q-systems $\mathbf{A}_\varphi$ induced from a Q-system $\mathbf{A}$ precisely give the Morita equivalence class of $\mathbf{A}$. However, inequivalent $\varphi$ may induce equivalent Q-systems $\mathbf{A}_\varphi$: e.g., if $\mathbf{A} = (\mathrm{id}, 1, 1)$ is the trivial Q-system, then all invertible $\varphi$, hence $\bar{\varphi}\varphi = \mathrm{id}$, induce the trivial Q-system.

## 3.6 Bimodules

The identification Sect. 3.4 between standard modules (= left modules) of a Q-system $\mathbf{A}$ in $\mathscr{C} \subset \mathrm{End}_0(N)$ and homomorphisms $N \to M$ of the associated pair of algebras works exactly the same for standard right modules $\mathbf{m} = (\beta, m \in \mathrm{Hom}(\beta, \beta\theta))$

(satisfying the analogous relations with the reversed tensor product). The correspondence is then that every standard right module is of the form

$$(\beta = \varphi\iota, m = \varphi(v)),$$

where $\varphi : M \to N$ is a sub-homomorphism of $\beta\bar{\iota}$.

In particular, a Q-system **A** is also a standard right **A**-module $(\beta = \theta, m = x)$, and the corresponding homomorphism is $\varphi = \bar{\iota} : M \to N$.

By obvious generalizations of the arguments, one also treats bimodules. An $\mathbf{A}_1$-$\mathbf{A}_2$-**bimodule** between two Q-systems is a triple $\mathbf{m} = (\beta, m_1 \in \mathrm{Hom}(\beta, \theta_1\beta), m_2 \in \mathrm{Hom}(\beta, \beta\theta_2))$ such that $(\beta, m_1)$ is a left $\mathbf{A}_1$-module and $(\beta, m_2)$ is a right $\mathbf{A}_2$-module, and the left and right actions commute:

$$(1_{\theta_1} \times m_2) \circ m_1 = (m_1 \times 1_{\theta_2}) \circ m_2 :$$

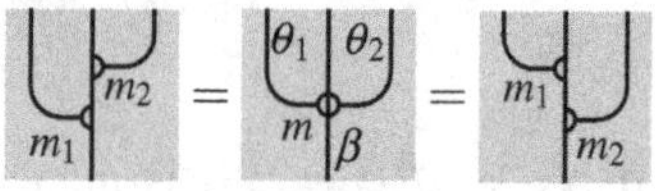


Equivalently, one may characterize the bimodule as a pair $\mathbf{m} = (\beta \in \mathscr{C}, m \in \mathrm{Hom}(\beta, \theta_1\beta\theta_2))$ satisfying

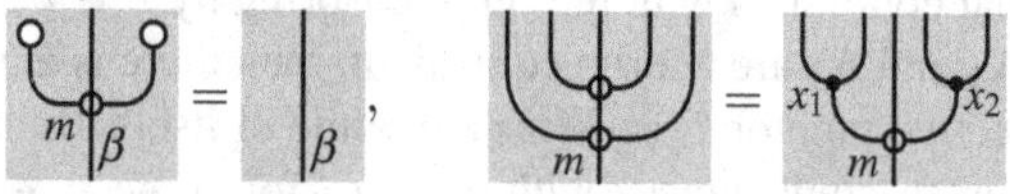


(3.6.1)

Then $(\beta, m_1) := (1_{\theta_1} \times 1_\beta \times w_2^*) \circ m$ is a left $\mathbf{A}_1$-module, $(\beta, m_2) := (w_1^* \times 1_\beta \times 1_{\theta_2}) \circ m$ is a right $\mathbf{A}_2$-module, and their actions commute.

A bimodule of a Q-system is called a **standard bimodule** if $m^*m$ is a multiple of $1_\beta$.

A Q-system **A** is also a standard **A**-**A**-bimodule $\mathbf{A} = (\beta = \theta, m = x^{(2)})$.

One proves the analogs of Lemmas 3.22 and 3.23, Propositions 3.24 and 3.25 and Corollary 3.26 in more or less exactly the same way (replacing the left trace in Proposition 3.25 by the right trace for the right module action):

**Proposition 3.32** (i) *Every bimodule is equivalent to a standard bimodule. The normalization of a standard bimodule is* $m^*m = d_{\mathbf{A}_1} d_{\mathbf{A}_2} \cdot 1_\beta$. *The adjoint of a bimodule is obtained by Frobenius reciprocity.*

(ii) [16] *Every standard bimodule is unitarily equivalent to a bimodule of the form* $\beta = \bar{\iota}_1\varphi\iota_2$ *where* $\varphi : M_2 \to M_1$ *is a sub-homomorphism of* $\iota_1\rho\bar{\iota}_2$ *for some* $\rho \in \mathscr{C} \subset \mathrm{End}_0(N)$ *(e.g.,* $\rho = \beta$*), and*

$$b = 1_{\bar{\iota}_1} \times v_1 \times 1_\varphi \times v_2 \times 1_{\iota_2} :$$

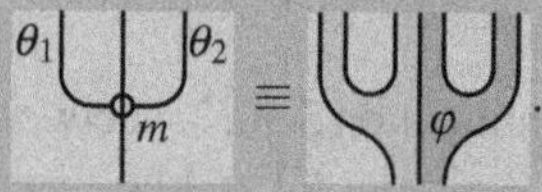

A morphism between two bimodules is an element $t \in \mathrm{Hom}(\beta, \beta')$ satisfying

$(1_{\theta_1} \times t \times 1_{\theta_2}) \circ m = m' \circ t$ : 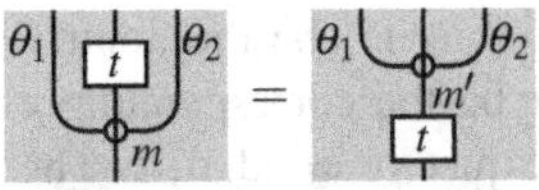

. (3.6.2)

**Proposition 3.33** *Every morphism t between two standard* $\mathbf{A}_1$*-*$\mathbf{A}_2$*-bimodules* $(\bar{\iota}_1\varphi\iota_2, 1_{\bar{\iota}_1} \times v_1 \times 1_\varphi \times v_2 \times 1_{\iota_2})$ *and* $(\bar{\iota}_1\varphi'\iota_2, 1_{\bar{\iota}_1} \times v_1 \times 1'_\varphi \times v_2 \times 1_{\iota_2})$ *is of the form* $t = 1_{\bar{\iota}_1} \times s \times 1_{\iota_2}$ *where* $s \in \mathrm{Hom}(\varphi, \varphi')$*. This establishes a bijective functor between the category of* $\mathbf{A}_1$*-*$\mathbf{A}_2$*-bimodules and the full subcategory of* $\mathrm{Hom}(M_2, M_1)$ *whose objects are the homomorphisms* $\prec \iota_1\rho\bar{\iota}_2$, $\rho \in \mathscr{C}$.

Again, the homomorphism associated with a standard bimodule $\mathbf{m} = (\beta, m)$ can be computed. Namely, the formula for $m$ implies that $\bar{\iota}_1\varphi(v_2) = w_1^* m$ (corresponding to $\mathbf{m}$ as a right $\mathbf{A}_2$-module). Hence $\varphi(\iota_2(n)v_2) = \iota_1(k)v_1$ implies $\beta(n)w_1^* m = \theta_1(k)x_1$, hence $k = w_1^*\beta(n)w_1^* m$:

$$\varphi(\iota_2(n)v_2) = \iota_1(w_1^*\beta(n)w_1^* m)v_1. \tag{3.6.3}$$

In particular, $\varphi(v_2) = \iota_1(w_1^* w_1^* m)v_1$.

The homomorphism associated with $\mathbf{A}$ as an $\mathbf{A}$-$\mathbf{A}$-bimodule is $\varphi = \mathrm{id}_M : M \to M$.

If $\varphi = \iota_1\rho\bar{\iota}_2$ (which is in general reducible), hence $\beta = \theta_1\rho\theta_2$ and $m = x_1\theta_1\rho(x_2)$, this simplifies to $\varphi(\iota_2(n)) = \iota_1(\rho\theta_2(n))$ and $\varphi(v_2) = \iota_1(\rho(x_2))$. Thus, $\varphi : M_2 \to M_1$ happens to take values in $\iota_1(N) \subset M_1$. This property is, however, not intrinsic, as it is not stable under unitary equivalence in the target algebra $M_1$. Also, the decomposition of $\varphi$ into irreducibles (which are unique only up to unitary equivalence within $M_1$) depends on the choice of isometries $s$, so that $\varphi_s = s^*\varphi(\cdot)s$. These may or may not be chosen in $\iota_1(N)$. As the Example 3.35 shows, there may be good reasons to choose the homomorphisms *not* to take values in $\iota_1(N)$.

Also the analog of Proposition 3.31 holds for bimodules, again with the same proof as for modules:

**Proposition 3.34** *There is a bijective functor between the category of* $\mathbf{A}_1$*-*$\mathbf{A}_2$*-bimodules and the category of* $\mathbf{A}'_1$*-*$\mathbf{A}'_2$*-bimodules, if and only if* $\mathbf{A}'_1$ *is induced from* $\mathbf{A}_1$ *by a standard module of* $\mathbf{A}_1$ *(i.e.,* $\iota_1 \prec \iota_2\rho$, $\rho \in \mathscr{C}$*), and* $\mathbf{A}'_2$ *is induced from* $\mathbf{A}_2$ *by a standard module of* $\mathbf{A}_2$.

In particular, the category of bimodules between a pair of Q-systems depends only on the Morita equivalence classes of the latter.

*Example 3.35* (Bimodules in the Ising category) Let $\mathbf{A} = (\theta = \sigma^2, w = 2^{\frac{1}{4}}r, x = 2^{-\frac{1}{4}}(r+t))$ be the nontrivial Q-system as in Example 3.14 and $M = N \vee \psi$ be the corresponding extension of $N$. The irreducible id-id-bimodules are just $\rho = \mathrm{id}, \sigma, \tau$. The $\mathbf{A}$-id-bimodules are the same as the modules of $\mathbf{A}$, Example 3.28.

The id-$\mathbf{A}$-bimodules arising from $\varphi = \rho\bar{\iota}$ are: $\mathbf{m}_\rho = (\beta = \rho\theta = \rho\sigma^2, m = \rho(x))$, where $x = 2^{-\frac{1}{4}}(r+t)$. Thus, $\varphi$ maps $n \in N$ to $\rho\sigma^2(n)$ and $v$ to $\rho(x)$.

(i) $\rho = \mathrm{id}$ and $\rho = \tau$: These are the same bimodules, because $\tau\sigma = \sigma$ and $\tau(x) = x$. One finds $\varphi : n \mapsto \sigma^2(n), \psi \mapsto rt^* + tr^*$.

(ii) $\rho = \sigma$: $\beta = \sigma^3$, $m = \sigma(x) = 2^{-\frac{1}{4}}\sigma(r+t) = 2^{-\frac{3}{4}}(r+t+(r-t)u)$. $\varphi : n \mapsto \sigma^3(n), \psi \mapsto rur^* - tut^*$.
The latter homomorphism $\varphi = \sigma\bar{\iota}$ is reducible, with projections $rr^*$ and $tt^*$ in the commutant. Then $\varphi_1 = r^*\varphi(\cdot)r$ and $\varphi_2 = t^*\varphi(\cdot)t$ give rise to

(ii.1) $\beta_1 = \sigma, m_1 = 2^{\frac{1}{4}}r, \varphi_1 : n \mapsto \sigma(n), \psi \mapsto u$.

(ii.2) $\beta_2 = \tau\sigma = \sigma, m_2 = 2^{\frac{1}{4}}t, \varphi_2 : n \mapsto \sigma(n), \psi \mapsto -u$. One has $\varphi_2 = \varphi_1 \circ \alpha = \tau \circ \varphi_1$.

The $\mathbf{A}$-$\mathbf{A}$-bimodules arising from $\varphi = \iota\rho\bar{\iota}$ are: $\mathbf{m}_\rho = (\beta = \theta\rho\theta = \sigma^2\rho\sigma^2, m = x\theta\rho(x))$. Thus $\varphi$ maps $n$ to $\rho\sigma^2(n)$ and $v$ to $\rho(x)$.

(i) $\rho = \mathrm{id}$ and $\rho = \tau$ are again the same bimodule. $\varphi : n \mapsto \sigma^2(n), \psi \mapsto rt^*+tr^*$. $\varphi$ is reducible with projections $\frac{1}{2}(1 \pm \psi) = s_\pm s_\pm^*, s_\pm = 2^{-\frac{1}{2}}(r \pm t\psi)$. This gives the irreducible components $\varphi_\pm : n \mapsto n, \psi \mapsto \pm\psi$, i.e., $\varphi_+ = \mathrm{id}$ and $\varphi_- = \alpha$.

(ii) $\rho = \sigma$. $\varphi : n \mapsto \sigma^3(n), \psi \mapsto \sigma(rt^* + tr^*) = rur^* - tut^*$. The commutant of $\varphi(M)$ contains $u$ and $\psi$. Thus $\varphi$ is the direct sum of two equivalent components, $[\varphi] = [\varphi'] \oplus [\varphi']$. Choosing projections $rr^*$ and $tt^*$ to compute $\varphi_1 = r^*\varphi(\cdot)r$ and $\varphi_2 = t^*\varphi(\cdot)t$, one has $\varphi_1 : n \mapsto \sigma(n), \psi \mapsto u$ and $\varphi_2 : n \mapsto \sigma(n), \psi \mapsto -u$. These are equivalent to each other by $\psi \in \mathrm{Hom}(\varphi_1, \varphi_2)$. They are also equivalent to the $\alpha$-inductions $\alpha_\sigma^\pm$ (cf. Sect. 4.6) by $U_\pm = 2^{-\frac{1}{2}}(1 \pm i\psi)$; with the former choice, $\varphi_i$ take values in $\iota_1(N)$, while $\mathrm{Ad}_{U_\pm}\varphi_1 = \mathrm{Ad}_{U_\mp}\varphi_2 = \alpha_\sigma^\pm$ don't.)

## 3.7 Tensor Product of Bimodules

The tensor product of bimodules is defined as follows. If $\mathbf{m}_1 = (\beta_1, m_1)$ is an **A**-**B**-bimodule and $\mathbf{m}_2 = (\beta_2, m_2)$ is an **B**-**C**-bimodule, then

$\widehat{m} =$ 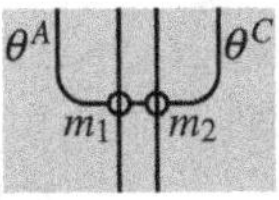 $\in \mathrm{Hom}(\beta_1\beta_2, \theta^{\mathrm{A}}\beta_1\beta_2\theta^{\mathrm{C}})$



satisfies the representation property of an **A**-**C**-bimodule, but the unit property fails. Instead, we have

**Lemma 3.36** *The intertwiner*

$p := d_{\mathrm{B}}^{-1} \cdot$ 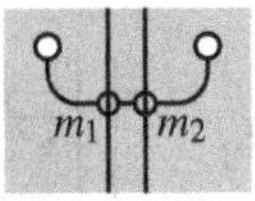 $\equiv d_{\mathrm{B}}^{-1} \cdot$ 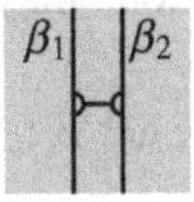 $\in \mathrm{Hom}(\beta_1\beta_2, \beta_1\beta_2)$



*is a projection, and satisfies*

$(1_{\theta^{\mathrm{A}}} \times p \times 1_{\theta^{\mathrm{C}}}) \circ \widehat{m} = \widehat{m} = \widehat{m} \circ p: \; d_{\mathrm{B}}^{-1} \cdot$ 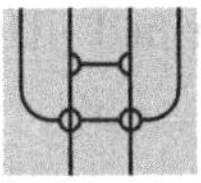 $=$ 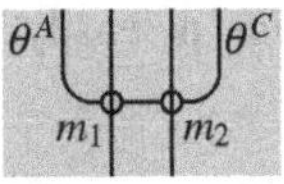 $= d_{\mathrm{B}}^{-1} \cdot$ 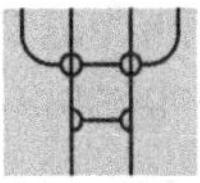 . (3.7.1)



*Proof* Idempotency of $p$ follows from the relation Eq. (3.7.1). Self-adjointness of $p$ follows from Lemma 3.23. To prove Eq. (3.7.1), we use the representation property, e.g.,

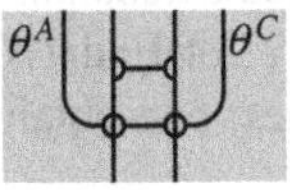 $=$ 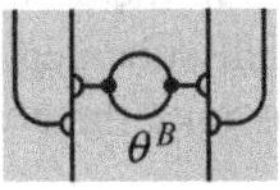 $= d_{\mathrm{B}} \cdot$ 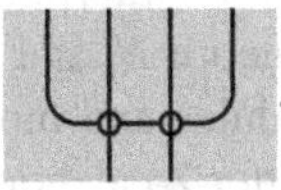 .



□

Then the bimodule tensor product is defined as the range of the projection $p$:

**Definition 3.37** Let $\mathbf{m}_1 = (\beta_1, m_1)$ be an **A**-**B**-bimodule and $\mathbf{m}_2 = (\beta_2, m_2)$ a **B**-**C**-bimodule. Choose an isometry $s \in N$ such that $ss^* = p$ and put $\beta(\cdot) := s^*\beta_1\beta_2(\cdot)s$ the range of $p$ in $\beta_1\beta_2$. Then the **bimodule tensor product**

$$\mathbf{m}_1 \otimes_{\mathbf{B}} \mathbf{m}_2 = (\beta, m),$$

$m := d_{\mathbf{B}}^{-1} \cdot (1_{\theta^{A}} \times s^* \times 1_{\theta^{C}}) \circ \widehat{m} \circ s = d_{\mathbf{B}}^{-1} \cdot$ 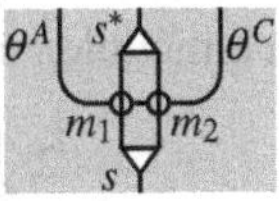 $\in \mathrm{Hom}(\beta, \theta^{\mathbf{A}}\beta\theta^{\mathbf{C}})$



is an **A**-**C**-bimodule.

**Proposition 3.38** *Under the correspondence Proposition* 3.32(ii), *the bimodule tensor product* $\mathbf{m}_1 \otimes_{\mathbf{B}} \mathbf{m}_2$ *corresponds to the composition of homomorphisms* $\varphi_1 \circ \varphi_2 : M^{\mathbf{C}} \to M^{\mathbf{A}}$.

*Proof* Using Proposition 3.32(ii), one computes

$$p = d_{\mathbf{B}}^{-1} \cdot 1_{\bar{\iota}^{\mathrm{A}} \circ \varphi_1} \times w^{\mathbf{B}} w^{\mathbf{B}*} \times 1_{\varphi_2 \circ \iota^{\mathrm{C}}} = d_{\mathbf{B}}^{-1} \cdot$$ 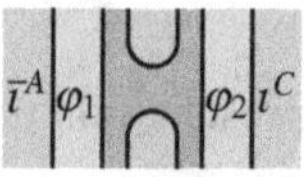

,

hence (up to unitary equivalence) one may choose

$$s = d_{\mathbf{B}}^{-\frac{1}{2}} \cdot 1_{\bar{\iota}^{\mathrm{A}} \circ \varphi_1} \times w^{\mathbf{B}} \times 1_{\varphi_2 \circ \iota^{\mathrm{C}}} \equiv d_{\mathbf{B}}^{-\frac{1}{2}} \cdot$$ 

.

With this, the claim is easily verified. The proper normalization is fixed by Proposition 3.32(i). □

In particular, we have equipped the category of **A**-**A**-bimodules with the structure of a tensor category, such that the tensor product corresponds to the composition of the corresponding endomorphisms in $\mathrm{End}_0(M)$. By admitting bimodules between different Q-systems $\mathbf{A}_i$, one arrives naturally at a (non-strict) bicategory (with 1-objects $\mathbf{A}_i$, 1-morphisms the bimodules and 2-morphisms the bimodule morphisms), corresponding to homomorphisms among the associated extensions $M_i$. Fixing the von Neumann algebra $N$ and some full subcategory $\mathscr{C}$ of $\mathrm{End}_0(N)$ in which the Q-systems, bimodules and morphisms take their values, one obtains a full sub-2-category of the latter 2-category.

In the tensor category of **A**-**A**-bimodules, the bimodule **A** is the tensor unit. Correspondingly, this category is simple iff **A** is irreducible as a **A**-**A**-bimodule. The following Lemma characterizes the self-intertwiners of **A**:

**Lemma 3.39** $t \in \mathrm{Hom}(\theta, \theta)$ *is a self-morphism of* **A** *as left* (*right*) **A**-*module if and only if t satisfies the first* (*second*) *of Eq.* (3.1.5). $t \in \mathrm{Hom}(\theta, \theta)$ *is a self-morphism of* **A** *as an* **A**-**A**-*bimodule if and only if* $t \in \mathrm{Hom}_0(\theta, \theta)$.

*Proof* The first statement is just the definition of morphisms. We prove only the last statement. "If": obvious. "Only if": by applying the unit relation in several ways to the defining property of a bimodule morphism

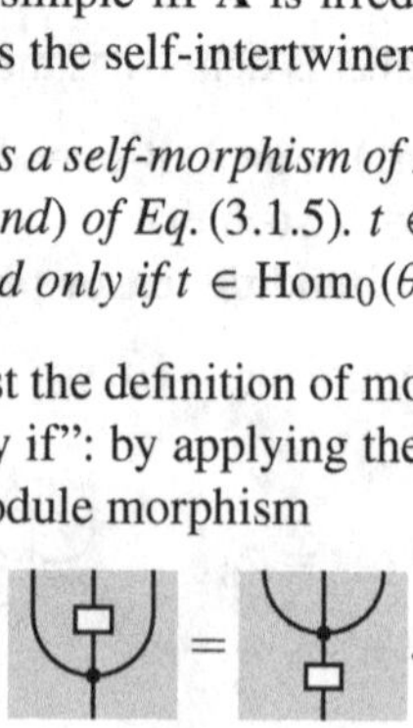
.

□

**Corollary 3.40** *The following are equivalent.*

(i) *A Q-system* **A** *is simple.*
(ii) *The corresponding extension $N \subset M$ is a factor.*
(iii) **A** *is irreducible as an* **A**-**A**-*bimodule.*
(iv) *The tensor category of* **A**-**A**-*bimodules is simple.*

Notice that (i) ⇔ (ii) is our Definition 3.15 of a simple Q-system. (iii) ⇔ (iv) is the definition of a simple tensor category. Thus, Corollary 3.40 states the equivalence of our Definition 3.15 of simplicity with the standard definition, which is given by the condition (iv).

*Proof* It suffices to prove (ii) ⇔ (iii). The endomorphism $\varphi : M \to M$ corresponding to the bimodule **A** according to Proposition 3.24, is $\varphi = \mathrm{id}_M$. Then, by Proposition 3.33, every self-intertwiner of **A** as an **A**-**A**-bimodule is of the form $t = 1_{\bar{\iota}} \times s \times 1_\iota \in \mathrm{Hom}(\theta, \theta)$, where $s \in \mathrm{Hom}(\mathrm{id}_M, \mathrm{id}_M)$. But $\mathrm{Hom}(\mathrm{id}_M, \mathrm{id}_M)$ is the same as the centre $M' \cap M$. □

For later use, we mention

**Lemma 3.41** *If* $\mathbf{m}_i = (\beta_i, m_i)$ $(i = 1, 2)$ *are* **A**-**B**-*bimodules, and* $t \in \mathrm{Hom}(\beta_1, \beta_2)$, *then*

$$S := m_2^* \circ (1_{\theta_\mathbf{A}} \times t \times 1_{\theta_\mathbf{B}}) \circ m_1 =$$ 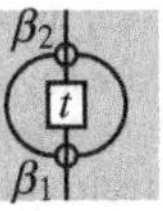

$$\in \mathrm{Hom}(\beta_1, \beta_2)$$

*is a bimodule morphism* : $\mathbf{m}_1 \to \mathbf{m}_2$.

The proof is rather easy in terms of the defining properties of modules and module intertwiners, and actually becomes trivial if one uses Proposition 3.32: namely $S \in 1_{\bar{\iota}_\mathbf{A}} \times \mathrm{Hom}(\varphi_1, \varphi_2) \times 1_{\iota_\mathbf{B}}$.

This Lemma implies that if the two bimodules are irreducible and inequivalent, then every intertwiner $S$ obtained in this way must be trivial. E.g., if $\mathbf{m}_2$ is trivial **A**-**A**-bimodule $(\theta, x^{(2)})$, and $\mathbf{m} = (\beta, m)$ is any nontrivial irreducible **A**-**A**-bimodule, then $x^* \circ (1_\theta \times s^* \times 1_\theta) \circ m =$  $= 0$ for every $s \in \mathrm{Hom}(\mathrm{id}, \beta)$. This is a special case of the Lemma (with $t = w \circ s^* \in \mathrm{Hom}(\beta, \theta)$), that we shall make use of in Sect. 4.12.

## References


1. S. Doplicher, R. Haag, J.E. Roberts, Local observables and particle statistics. I. Commun. Math. Phys. **23**, 199–230 (1971)
2. D. Guido, R. Longo, The conformal spin and statistics theorem. Commun. Math. Phys. **181**, 11–35 (1996)
3. R. Longo, Index of subfactors and statistics of quantum fields I. Commun. Math. Phys. **126**, 217–247 (1989)
4. K. Fredenhagen, K.-H. Rehren, B. Schroer, Superselection sectors with braid group statistics and exchange algebras I. Commun. Math. Phys. **125**, 201–226 (1989)
5. J. Fröhlich, J. Fuchs, I. Runkel, C. Schweigert, Correspondences of ribbon categories. Ann. Math. **199**, 192–329 (2006)
6. J. Fuchs, I. Runkel, C. Schweigert, TFT construction of RCFT correlators I: partition functions. Nucl. Phys. B **646**, 353–497 (2002)
7. R. Longo, J.E. Roberts, A theory of dimension. K-Theory **11**, 103–159 (1997). (notably Chaps. 3 and 4)
8. M. Izumi, H. Kosaki, On a subfactor analogue of the second cohomology. Rev. Math. Phys. **14**, 733–737 (2002)
9. R. Longo, A duality for Hopf algebras and for subfactors. Commun. Math. Phys. **159**, 133–150 (1994)
10. M. Bischoff, Y. Kawahigashi, R. Longo, K.-H. Rehren, Phase boundaries in algebraic conformal QFT. arXiv:1405.7863
11. V.F.R. Jones, Index for subfactors. Invent. Math. **72**, 1–25 (1983)
12. V.F.R. Jones, The planar algebra of a bipartite graph, in *Knots in Hellas '98*, Series Knots Everything, vol. 24 (World Scientific, River Edge, 2000), pp. 94–117
13. R. Longo, K.-H. Rehren, Nets of subfactors. Rev. Math. Phys. **7**, 567–597 (1995)
14. M. Müger, From subfactors to categories and topology I. Frobenius algebras in and Morita equivalence of tensor categories. J. Pure Appl. Algebra **180**, 81–157 (2003)
15. M. Müger, From subfactors to categories and topology II. The quantum double of tensor categories and subfactors. J. Pure Appl. Algebra **180**, 159–219 (2003)
16. D. Evans, P. Pinto, Subfactor realizations of modular invariants. Commun. Math. Phys. **237**, 309–363 (2003)
17. V. Ostrik, Module categories, weak Hopf algebras and modular invariants. Transform. Groups **8**, 177–206 (2003)

# Chapter 4
# Q-System Calculus

**Abstract** We introduce operations with Q-systems and clarify their meaning in terms of the corresponding extensions $N \subset M$. We identify three different types of reduced Q-systems, corresponding to the central decomposition and the irreducible decomposition of the extension, and to intermediate extensions. In braided tensor categories, the centre, the full centre, and the braided product of Q-systems are defined. The main classification result is the computation of the central decomposition of the braided product of the full centres of two Q-systems in a modular C* tensor category.

Throughout this section, $N$ is an infinite factor, and $\mathscr{C} \subset \mathrm{End}_0(N)$ with properties as specified in Chap. 3.

Q-systems in $\mathscr{C}$ can be decomposed in several distinct ways. In the first four sections, we discuss various decompositions in turn, and characterize them in terms of suitable projections in the underlying category $\mathscr{C}$.

In the remainder of this section, we discuss Q-systems in braided C* tensor categories, introduce various operations with Q-systems (the centres, the braided products and the full centre), and compute the central decomposition of the extension corresponding to the braided product of two full centres. The latter is motivated because this decomposition gives the irreducible boundary conditions for phase boundaries in local QFT [1].

## 4.1 Reduced Q-Systems

Let $(\theta, w, x)$ be a Q-system describing the extension $N \subset M$. When the multiplicity $\dim\mathrm{Hom}(\mathrm{id}_N, \theta) = \dim\mathrm{Hom}(\iota, \iota)$ of $\mathrm{id}_N$ in $\theta$ is one, then the extension is irreducible ($N' \cap M = \mathbb{C}\mathbf{1}$), and in particular $M$ is automatically a factor. When the multiplicity is larger than one, then $M$ may or may not be a factor.

Let $e$ be a nontrivial projection in $\mathrm{Hom}(\iota, \iota)$. If $M$ is a factor, then one can write $e = tt^*$ with an isometry $t \in M$, define a sub-homomorphism $\iota_e \prec \iota$ by $\iota_e(\cdot) = t^*\iota(\cdot)t$, and arrive at a decomposition $[\iota] = [\iota_e] \oplus [\iota_{1-e}]$, cf. Corollary 4.10, where we shall characterize this decomposition in terms of certain projections in $\mathrm{Hom}(\theta, \theta)$.

 
M. Bischoff et al., *Tensor Categories and Endomorphisms of von Neumann Algebras*, SpringerBriefs in Mathematical Physics, DOI 10.1007/978-3-319-14301-9_4

In contrast, if $M$ is not a factor and $e \neq 1$ belongs to the centre of $M$, such isometries $t$ do not exist in $M$. Namely, $t \in M$ and $tt^* \in M'$ would imply $e = et^*t = t^*et = \mathbf{1}_M$.

One should therefore first perform a central decomposition of $M$ into factors $M_e = eM$ by the minimal central projections, and compute the reduced Q-systems (cf. Sect. 4.2) for the subfactors $N \simeq Ne \subset Me$. Each of these may still be reducible, and can be further reduced by decomposing $[\iota] = \bigoplus_e [\iota_e]$, as before.

Finally, we also discuss in Sect. 4.4 the multiplicative "splitting" decomposition of $\iota$, when there is an intermediate subfactor $N \subset L \subset M$, so that $\iota = \iota_2 \circ \iota_1$. Whether an intermediate subfactor exists is independent of reducibility of the subfactor, e.g., $[\iota] = [\mathrm{id}_N] \oplus [\mathrm{id}_N]$ is reducible but does not admit an intermediate subfactor, whereas $\iota_1 \otimes \iota_2 = (\iota_1 \otimes \mathrm{id}_{N_2}) \circ (\mathrm{id}_{N_1} \otimes \iota_2) : N_1 \otimes N_2 \to M_1 \otimes M_2$ is an irreducible homomorphism (if $\iota_i$ are) but admits intermediate factors $N_1 \otimes M_2$ and $M_1 \otimes N_2$. This example also shows that the splitting cannot be expected to be unique. Also, even if both $N$ and $M$ are factors, the intermediate algebra $L$ need not be a factor, as the example $N \subset N \oplus N \subset \mathrm{Mat}_2(N)$ shows.

Although the three decompositions of a Q-system $(\theta, w, x)$ are of quite different nature, they all come with a projection $P \in \mathrm{Hom}(\theta, \theta)$ satisfying

$$(P \times P) \circ x = (P \times 1_\theta) \circ x \circ P = (1_\theta \times P) \circ x \circ P$$

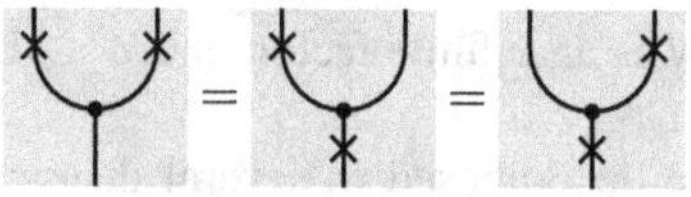

(4.1.1)

("of three projection, any one is redundant", or "any two projections imply the third"), and in each case different further properties. One easily proves:

**Lemma 4.1** *Equation* (4.1.1) *alone implies*

(i) *The triple*

$$\theta_P := S^* \theta(\cdot) S, \quad \widetilde{w}_P := S^* \circ w, \quad \widetilde{x}_P := (S^* \times S^*) \circ x \circ S$$

*is a* $C^*$ *Frobenius algebra, where* $S \in N$ *is any isometry such that* $SS^* = P$, *i.e.,* $\theta_P \prec \theta$.

(ii) $n_P := \widetilde{x}_P^* \widetilde{x}_P$ *is a multiple of* $1_{\theta_P}$ *if and only if* $x^* \circ (P \times P) \circ x$ *is a multiple of* $P$.

(The notation emphasizes that the unitary equivalence class of $(\theta_P, \widetilde{w}_P, \widetilde{x}_P)$ depends on $P$, but not on the choice of the isometry $S$.)

*Proof* (i) The unit property, associativity and Frobenius property follow from the corresponding properties of $(\theta, w, x)$ by "eliminating" projections using Eq. (4.1.1) and $PS = S$.

(ii) "If" is obvious. Conversely, $\widetilde{x}_P^* \widetilde{x}_P = \mu \cdot 1_{\theta_P}$ implies that $P \circ x^* \circ (P \times P) \circ x \circ P = \mu \cdot P$. By Eq. (4.1.1), this equals $x^* \circ (P \times P) \circ x$. □

However, the property in (ii) may fail, in which case the C* Frobenius algebra fails to be special (and hence to be standard). Then, by Corollary 3.6, one can define the equivalent special C* Frobenius algebra $(\theta_P, \widehat{w}_P, \widehat{x}_P)$ with

$$\widehat{w}_P := n_P^{\frac{1}{2}} S^* \circ w, \qquad \widehat{x}_P := (n_P^{-\frac{1}{2}} S^* \times n_P^{-\frac{1}{2}} S^*) \circ x \circ S n_P^{\frac{1}{2}}$$

with the "normalization intertwiner" $n_P = \widetilde{x}_P^* \widetilde{x}_P = S^* X^*(P \times P) X S \in \mathrm{Hom}_0(\theta_P, \theta_P)$, such that $\widehat{x}_P^* \widehat{x}_P = 1_{\theta_P}$.

$\widehat{w}_P \in \mathrm{Hom}(\mathrm{id}, \theta_P)$ is automatically a multiple of an isometry, and

$$\widehat{w}_P^* \widehat{w}_P = w^* \circ x^* \circ (P \times P) \circ x \circ w = r^* \circ (P \times P) \circ r.$$

**Corollary 4.2** *The appropriately rescaled triple*

$$\theta_P = S^* \theta(\cdot) S,$$

$$w_P = \dim(\theta_P)^{-\frac{1}{4}} \cdot n_P^{\frac{1}{2}} S^* \circ w, \tag{4.1.2}$$

$$x_P = \dim(\theta_P)^{\frac{1}{4}} \cdot (n_P^{-\frac{1}{2}} S^* \times n_P^{-\frac{1}{2}} S^*) \circ x \circ S n_P^{\frac{1}{2}}$$

*is a standard C* Frobenius algebra, i.e., a Q-system, called the* **reduced Q-system**, *if and only if $\widehat{w}_P$ has the correct normalization $\widehat{w}_P^* \widehat{w}_P = r^* \circ (P \times P) \circ r \stackrel{!}{=} \dim(\theta_P)$.*

In each of the three decompositions discussed in the subsequent sections, further properties of the characterizing projections beyond Eq. (4.1.1) will indeed ensure the correct normalization as required in Corollary 4.2.

## 4.2 Central Decomposition of Q-Systems

In this section, we shall characterize decompositions of $\iota : N \to M$ as a direct sum

$$\iota = \iota_1 \oplus \iota_2$$

when $M = M_1 \oplus M_2$ is not a factor, $\iota_i : N \to M_i$.

Let $N \subset M$ an inclusion of von Neumann algebras, where $N$ is an infinite factor, and $M$ is properly infinite with a finite centre. Every central projection $e \in M' \cap M$ gives rise to an inclusion $eN \subset eM$, where $eN$ is canonically isomorphic to $N$. (Recall also the characterization of such projections as $e = \iota(q)v$ with $q \in \mathrm{Hom}(\theta, \mathrm{id}_N)$, given in Lemma 3.16.) If $e$ is minimal, $eN \subset eM$ is a subfactor.

We want to characterize the corresponding embeddings $eN \to eM$ in terms of reduced Q-systems Eq. (4.1.2). Our starting observation is that $P := \bar{\iota}(e) \in N$ is a projection in $\mathrm{Hom}_0(\theta, \theta)$:

$$(1_\theta \times P) \circ x = x \circ P = (P \times 1_\theta) \circ x :$$

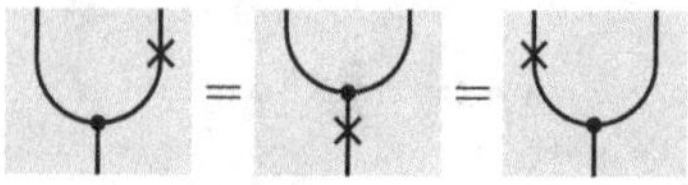

which is precisely Eq. (3.1.5). This follows immediately by applying $\bar{\iota}$ to the equations $\iota\bar{\iota}(e)v = ve$ (because $e \in M$) and $ev = ve$ (because $e \in M'$). We now show the converse:

**Proposition 4.3** *Let* $\mathbf{A} = (\theta, w, x)$ *be a Q-system defining the extension* $N \subset M$. *Let* $P \in \mathrm{Hom}(\theta, \theta)$ *be a projection satisfying Eq. (3.1.5), hence also Eq. (4.1.1). Then Eq. (4.1.2) with normalization intertwiner* $n_P = \sqrt{\dim(\theta)} \cdot 1_{\theta_p}$ *defines a reduced Q-system* $\mathbf{A}_P$. *The reduced Q-system corresponds to the extension* $eN \subset eM$ *where* $e \in M' \cap M$ *and* $\bar{\iota}(e) = P$.

*Along with* $P$, *also* $1 - P$ *satisfies Eq. (3.1.5).*

If $P$ is a *minimal* projection in $\mathrm{Hom}(\theta, \theta)$ with the stated properties, we will also refer to the reduced Q-system as a **factor Q-system** of $\mathbf{A}$.

*Proof* Let us first compute the normalizations. Let $S \in N$ be any isometry such that $P = SS^*$. Because $P \in \mathrm{Hom}_0(\theta, \theta)$, we have $n_P = S^*x^*(P \times P)xS = S^*Px^*xPS = \sqrt{\dim(\theta)} \cdot 1_\theta$, and $r^*(P \times P)r = r^*(1_\theta \times P)r = \mathrm{Tr}_\theta(P) = \dim(\theta_P)$ by Proposition 2.4. Hence, by Corollary 4.2, $\mathbf{A}_P = (\theta_P, w_P, x_P)$ is a reduced Q-system.

By Lemma 3.39, $P$ is a self-morphism of $\mathbf{A}$ considered as an $\mathbf{A}$-$\mathbf{A}$-bimodule. Hence, by Proposition 3.33, $P = 1_{\bar{\iota}} \times e \times 1_\iota$ with $e \in \mathrm{Hom}(\mathrm{id}_M, \mathrm{id}_M) = M' \cap M$. If $P$ is a projection, so is $e$. We claim that $n \mapsto en \equiv e\iota(n)$ is a *-isomorphism between $N$ and $eN$. Because $e$ is a central projection, the *-homomorphism property is obvious, and so is surjectivity. Injectivity follows because $e\iota(n) = 0$ implies $P\theta(n) = 0$, hence $\theta_P(n) := S^*\theta(n)S = 0$. Since $\theta_P$ is injective, $n = 0$.

We now define a conjugate $\bar{\iota}_P$ for the embedding $\iota_P : N \to eM, n \mapsto e\iota(n)$:

$$\bar{\iota}_P : em \mapsto S^*\bar{\iota}(m)S.$$

Then $\widetilde{w}_P := S^*w \in \mathrm{Hom}(\mathrm{id}_{eN}, \bar{\iota}_P\iota_P)$ and $\widetilde{v}_P := e\iota(S^*)v \in \mathrm{Hom}(\mathrm{id}_{eM}, \iota_P\bar{\iota}_P)$ are intertwiners:

$$\begin{aligned}\bar{\iota}_P\iota_P(n)\widetilde{w}_P &= S^*\theta(n)SS^*w = S^*\theta(n)Pw = S^*P\theta(n)w \\ &= S^*\theta(n)w = S^*wn = \widetilde{w}_Pn,\end{aligned}$$

$$\iota_P\bar{\iota}_P(em)\widetilde{v}_P = e\iota(S^*)\iota\bar{\iota}(m)\iota(SS^*)v = e\iota(S^*P)\iota\bar{\iota}(m)v = e\iota(S^*)vm = \widetilde{v}_Pem,$$

because $\iota(SS^*) = \iota(P) = \iota\bar{\iota}(e)$ commutes with $= \iota\bar{\iota}(m)$. $\widetilde{w}_P$ and $\widetilde{v}_P$ solve the conjugacy relations Eq. (2.2.1):

$$\widetilde{w}_P^*\bar{\iota}_P(\widetilde{v}_P) = w^*SS^*\bar{\iota}(e\iota(S^*)v)S = w^*P\theta(S^*)xS = w^*\theta(S^*)xS$$
$$= S^*w^*xS = S^*S = \mathbf{1}_N,$$

because $P$ commutes with $\theta(S^*)$ and with $x$, and

$$\bar{\iota}[\iota_P(\widetilde{w}_P^*)\widetilde{v}_P] = \bar{\iota}(e\iota(w^*S)\iota(S^*)v) = P\theta(w^*P)x = P\theta(w^*)xP = P^2 = P = \bar{\iota}(e),$$

which implies $\iota_P(\widetilde{w}_P^*)\widetilde{v}_P = e = 1_{eM}$. Finally, $\widetilde{x}_P := \bar{\iota}_P(\widetilde{v}_P)$ equals $S^*\bar{\iota}(e)\theta(S^*)xS = (S^* \times S^*)xS$ because $\bar{\iota}(e) = P$. Thus, after the appropriate rescaling by $\dim(\theta_P)^{\mp\frac{1}{4}} \cdot n_P^{\pm\frac{1}{2}} = (\dim(\theta_P)/\dim(\theta))^{\mp\frac{1}{4}}$, the Q-system for $eN \subset eM$ coincides with the reduced Q-system $\mathbf{A}_P = (\theta_P, w_P, x_P)$.

The last statement is obvious by linearity. □

**Corollary 4.4** *If $1_\theta = \sum P_i$ is the partition of unity into minimal projections in* $\mathrm{Hom}(\theta, \theta)$ *satisfying Eq.* (3.1.5)*, then $(\theta, w, x)$ is the direct sum of simple Q-systems as in Eq.* (3.2.2)*. The corresponding partition $\mathbf{1}_M = \sum e_i$ gives the decomposition of $M' \cap M$ into minimal central projections, i.e., each simple extension $e_iN \subset e_iM$ is a representation of the extension $N \subset M$.*

## 4.3 Irreducible Decomposition of Q-Systems

In this section, we shall characterize decompositions of $\iota : N \to M$ as a direct sum of sectors

$$[\iota] = [\iota_1] \oplus [\iota_2], \qquad \text{i.e.,} \quad \iota(\cdot) = s_1\iota_1(\cdot)s_1^* + s_2\iota_2(\cdot)s_2^*$$

for infinite subfactors $N \subset M$.

Thus, let both $N$ and $M$ be factors, i.e., there are no nontrivial projections in $\mathrm{Hom}(\theta, \theta)$ satisfying Eq. (3.1.5).

If $\iota(N)' \cap M = \mathrm{Hom}(\iota, \iota)$ is nontrivial, then $\iota$ is a reducible homomorphism. If $e \in \iota(N)' \cap M$ is a projection, then there is an isometry $s \in M$ such that $ss^* = e$, and $\iota_s(n) = s^*\iota(n)s$ is a sub-homomorphism of $\iota$. Clearly $s \in \mathrm{Hom}(\iota_s, \iota)$.

**Lemma 4.5** *The homomorphism $\iota_s : N \to M$ is isomorphic to the embedding $eN \equiv Ne \subset eMe$, i.e., the identical map $\iota_e : eN \to eMe$.*

*Proof* We may write $eN \subset eMe$ as $ss^*\iota(N)ss^* \equiv s\iota_s(N)s^* \subset sMs^*$. The claim follows because the map $\mathrm{Ad}_s : M \to sMs^* \equiv eMe$ is an isomorphism. □

For $\iota_s \prec \iota$ one has a conjugate $\overline{\iota_s} \prec \overline{\iota}$, and an isometry $\overline{s} \in \mathrm{Hom}(\overline{\iota}_s, \overline{\iota})$. Then $e = ss^* \in \mathrm{Hom}(\iota, \iota) \subset M$ and $\overline{e} = \overline{s}\overline{s}^* \in \mathrm{Hom}(\overline{\iota}, \overline{\iota}) \subset N$ are projections such that

$$(\overline{e} \times 1_\iota) \circ w = (1_{\overline{\iota}} \times e) \circ w, \qquad (e \times 1_{\overline{\iota}}) \circ v = (1_\iota \times \overline{e}) \circ v, \tag{4.3.1}$$

and $w^* \circ (\overline{e} \times e) \circ w = \dim(\iota_s) \cdot \mathbf{1}_N$, $v^* \circ (e \times \overline{e}) \circ v = \dim(\iota_s) \cdot \mathbf{1}_M$, Then $p = 1_{\overline{\iota}} \times e$ and $\overline{p} = \overline{e} \times 1_\iota$ are a pair of commuting projections $\in \mathrm{Hom}(\theta, \theta)$ such that

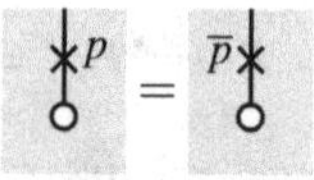

(4.3.2)

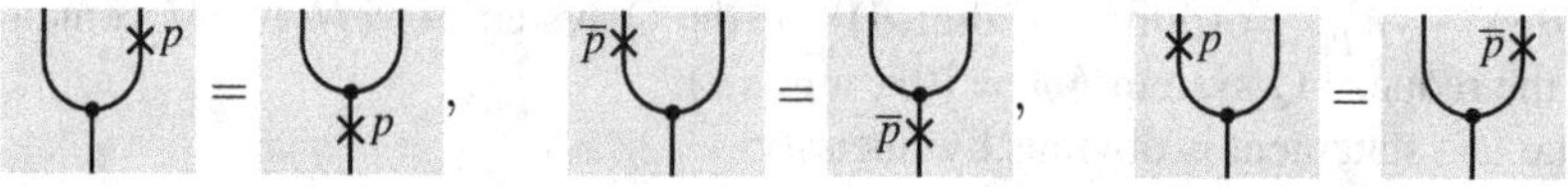

(4.3.3)

Conversely, if $\mathbf{A} = (\theta, w, x)$ is a simple Q-system, and $\dim \mathrm{Hom}(\mathrm{id}, \theta) > 1$, then by Frobenius reciprocity also $\dim \mathrm{Hom}(\iota, \iota) > 1$, hence $\iota$ is reducible. In order to decompose $\iota$, we want to characterize the projections in $\mathrm{Hom}(\iota, \iota)$ in terms of projections in $\mathrm{Hom}(\theta, \theta)$.

Instead of characterizing $\iota_1 \prec \iota$ by the pair of projections $p, \overline{p}$ satisfying Eqs. (4.3.2) and (4.3.3), we observe that either $p$ or $\overline{p}$ suffices: namely, from the third relation in (4.3.3), one can express $p$ in terms of $\overline{p}$, and vice versa:

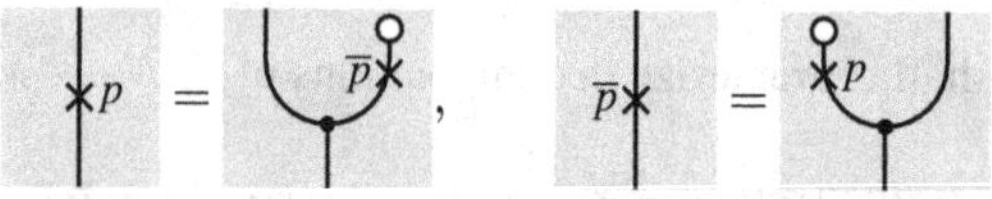

(4.3.4)

Expressing $p$ in terms of $\overline{p}$ as in Eq. (4.3.4), turns Eq. (4.3.2) into another relation for $\overline{p}$

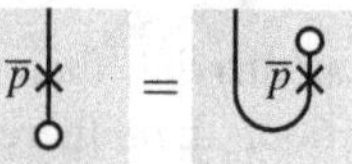

(4.3.5)

besides the second in (4.3.3), while the first is automatically satisfied. In the same way, expressing $\overline{p}$ in terms of $p$, turns Eq. (4.3.2) into another relation for $p$

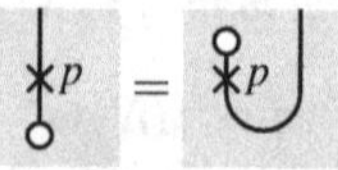

(4.3.6)

besides the first in (4.3.3).

**Lemma 4.6** *Let* $\mathbf{A} = (\theta, w, x)$ *be a simple Q-system. Let either* $p \in \mathrm{Hom}(\theta, \theta)$ *be a projection satisfying Eq.* (4.3.6) *and the first of Eq.* (4.3.3), *or* $\overline{p} \in \mathrm{Hom}(\theta, \theta)$ *a projection satisfying Eq.* (4.3.5) *and the second of Eq.* (4.3.3). *Defining* $\overline{p} :=$ 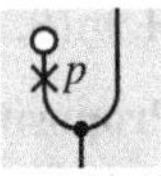 *in the first case, and* $p :=$ 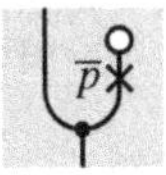 *in the second case, gives another projection such that* $p$ *and* $\overline{p}$ *satisfy the system Eqs.* (4.3.2) *and* (4.3.3).

*Proof* We first establish that $\overline{p}$ defined from $p$ is a projection:

$\overline{p}^* =$ 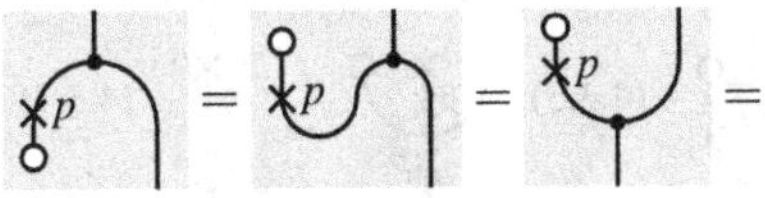 $= \overline{p},$

$\overline{p}^2 =$ 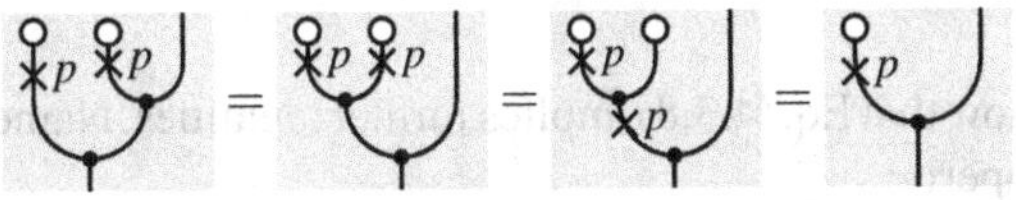 $= \overline{p},$

where we have used $p^* = p$ and the first defining property Eq. (4.3.6) of $p$, and $p^2 = p$ and the second defining property Eq. (4.3.3) of $p$, respectively.

That $\overline{p}$ satisfies Eq. (4.3.2), is an immediate consequence of the defining property Eq. (4.3.6) of $p$. The second of Eq. (4.3.3) is trivial by associativity. It remains to verify the last of Eq. (4.3.3):

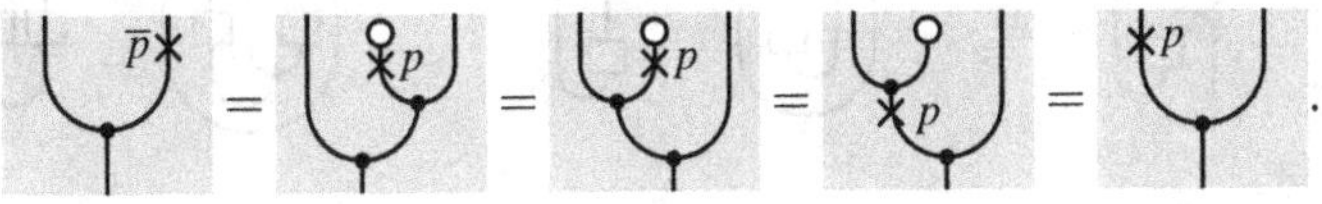

The properties of $p$ defined from $\overline{p}$ follow similarly. □

**Lemma 4.7** *If projections* $p \in \mathrm{Hom}(\theta, \theta)$ *and* $\overline{p} \in \mathrm{Hom}(\theta, \theta)$ *satisfy Eq.* (4.3.2) *and* (4.3.3), *then* $p$ *and* $\overline{p}$ *commute, and* $P = \overline{p}p$ *is a projection satisfying Eq.* (4.1.1).

*Proof* Using in turn Eq. (4.3.4), the second and the last relation of Eq. (4.3.3), one finds

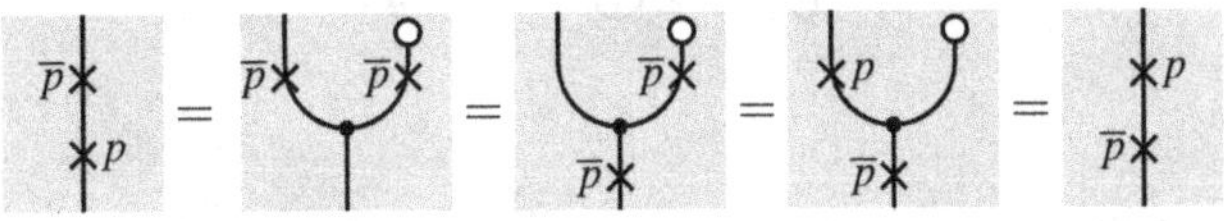

(4.3.7)

It follows that $P = \overline{p}p$ is a projection, and the relations Eq. (4.3.3) immediately imply Eq. (4.1.1) for $P$. □

Instead of characterizing either of the projections $p$ or $\overline{p} \in \mathrm{Hom}(\theta, \theta)$ as in Lemma 4.6, it is also possible to characterize directly the projection $P = \overline{p}p \in \mathrm{Hom}(\theta, \theta)$.

**Proposition 4.8** *A projection $P \in \mathrm{Hom}(\theta, \theta)$ is of the form $P = \overline{p}p$ with $p$ and $\overline{p}$ as in Lemma 4.6, if and only if $P$ satisfies*

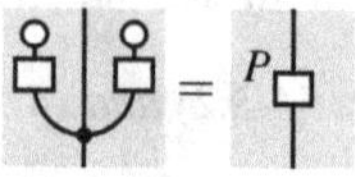


(4.3.8)

*Proof* "Only if": we show that $P = \overline{p}p$ satisfies Eq. (4.3.8):

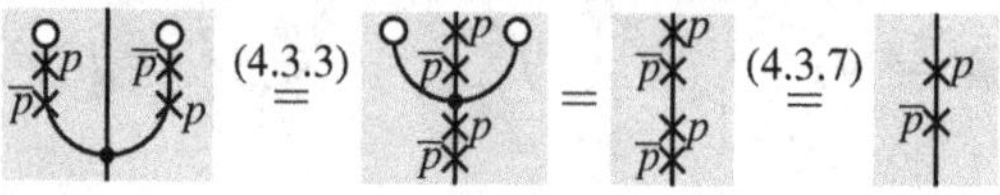


"If": We first show that Eq. (4.3.8) implies further identities. Namely, we obviously get by the unit property:

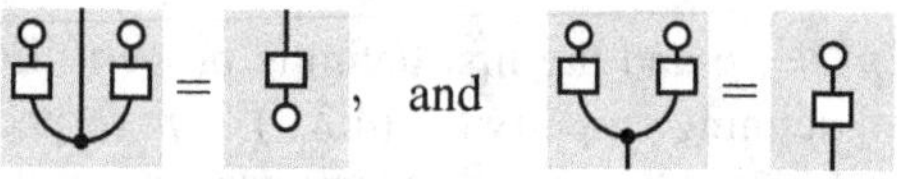


(4.3.9)

Moreover, by Proposition 2.6, we have

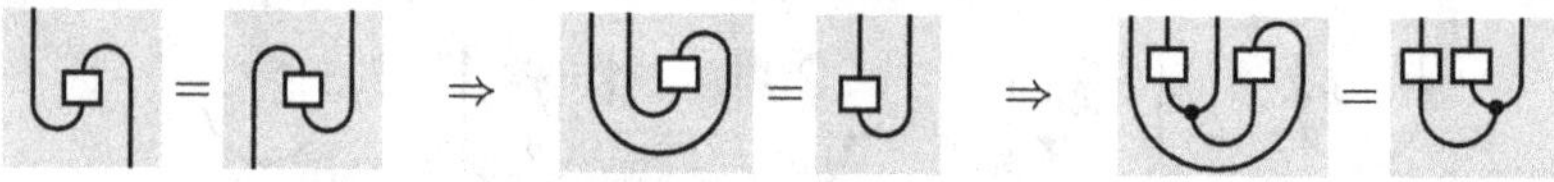

(4.3.10)

implying

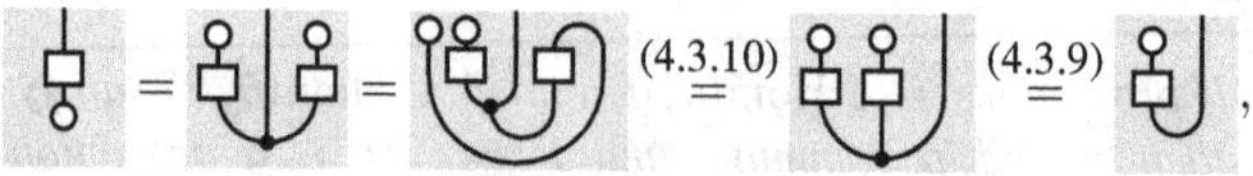


and similarly

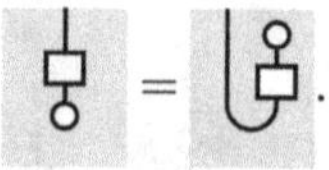

Now, given $P$, we define $\overline{p} := (w^* P \times 1_\theta) \circ x =$  and $p := (1_\theta \times w^* P) \circ x =$ 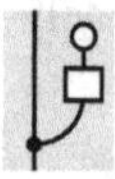. Then, obviously $P = \overline{p}p$, and $p$ and $\overline{p}$ satisfy the last of Eq. (4.3.3):

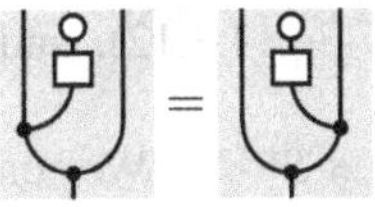

(4.3.11)

hence $p$ and $\overline{p}$ are related to each other by Eq. (4.3.4). Moreover, $\overline{p}$ obviously satisfies Eq. (4.3.5) and the second of Eq. (4.3.3) by associativity and the unit property. In view of Lemma 4.6, it remains to verify that $\overline{p}$ is a projection. Idempotency of $\overline{p}$ is just Eq. (4.3.9), and selfadjointness of follows from Eq. (4.3.11):

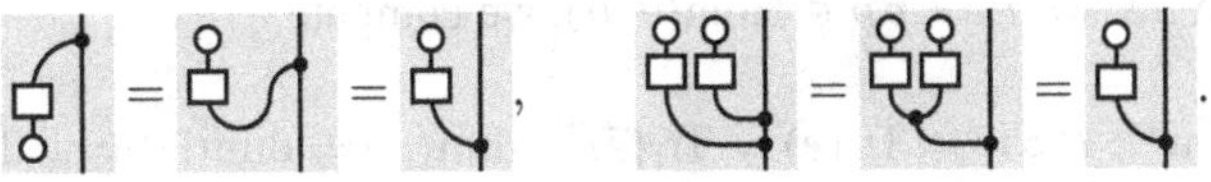

□

The following is the main result of this section.

**Proposition 4.9** *Let* $\mathbf{A} = (\theta, w, x)$ *be a simple Q-system, describing a subfactor* $N \subset M$. *Let either* $p$ *or* $\overline{p}$ *or* $P$ *be a projection in* $\mathrm{Hom}(\theta, \theta)$ *with properties as specified in Lemma* 4.6 *resp. Proposition* 4.8, *thus defining the respective other two projections. Then Eq.* (4.1.2) *with normalization factor* 1 *(i.e.,* $n_P = \sqrt{\dim(\theta_P)} \cdot 1_{\theta_p}$*) defines a reduced Q-system* $\mathbf{A}_P$. *The sub-Q-system* $\mathbf{A}_P$ *is associated with a homomorphism* $\iota_p \prec \iota$ *which is the range of* $p \in \mathrm{Hom}(\iota, \iota)$ *such that* $\iota$ *is a direct sum* $\iota_p \oplus \iota_{1-p}$.

We will also refer to this reduced Q-system as a sub-Q-system of **A**.

*Proof* From Lemma 4.6 we know that $p, \overline{p}$ satisfy the system Eqs. (4.3.2) and (4.3.3).

The first and second relations of Eq. (4.3.3) state that $p$ (resp. $\overline{p}$) are self-intertwiners of $(\theta, x)$ as a left (resp. right) **A**-module. By Proposition 3.25 for left and right modules, we conclude that $p = 1_{\overline{\iota}} \times e$ and $\overline{p} = \overline{e} \times 1_{\overline{\iota}}$ with projections $e \in \mathrm{Hom}(\iota, \iota)$, $\overline{e} \in \mathrm{Hom}(\overline{\iota}, \overline{\iota})$. In terms of the projections $e$ and $\overline{e}$, Eq. (4.3.2) and the last relation of Eq. (4.3.3) read

(4.3.12)

Because $M$ is a factor, one can write $e = ss^*$ with an isometry $s \in M$ and define $\iota_s = s^*\iota(\cdot)s \prec \iota$ as the range of $e$. Similarly, $\overline{\iota}_{\overline{s}} = \overline{s}^*\overline{\iota}(\cdot)\overline{s} \prec \overline{\iota}$ is the range of $\overline{e}$. Then $\iota_s$ and $\overline{\iota}_{\overline{s}} \prec \overline{\iota}$ are conjugate homomorphisms, because $w_P = (\overline{s}^* \times s^*) \circ w$, $v_P = (s^* \times \overline{s}^*) \circ v$ solve the conjugacy relations Eq. (2.2.1):

$(1_{\overline{\iota}_s} \times v_P^*) \circ (w_P \times 1_{\iota_s}) =$ 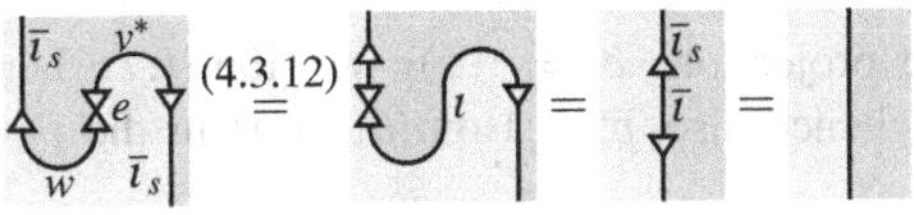 $= 1_{\overline{\iota}_s}$, where we have used the first of Eq. (4.3.12), and similarly $(1_{\iota_s} \times w_P^*) \circ (v_P \times 1_{\bar{\iota}_s}) = 1_{\iota_s}$.

Now let $S = \bar{s}^*\bar{\iota}(s)$, hence $w_P = S^* \circ w$. We compute

$$\begin{aligned}\bar{\iota}_s(v_P) &= \bar{s}^*\bar{\iota}[s^*\iota(\bar{s}^*)v]\bar{s} = \bar{s}^*\bar{\iota}[s^*\iota(\bar{s}^*\bar{\iota}(s^*))vs]\bar{s} \\ &= S^*\theta(\bar{s}^*\bar{\iota}(s^*))\bar{\iota}(vs)\bar{s} = (S^* \times S^*) \circ x \circ S =: x_P.\end{aligned}$$

It remains to show that $(\theta_P = \bar{\iota}_s\iota_s, w_P, x_P)$ is the reduced Q-system. Clearly, $\theta_P = S^*\theta(\cdot)S$. With $SS^* = P = \bar{p}p \in \mathrm{Hom}(\theta, \theta)$, we compute

$$w_P^* w_P = w^* \circ P \circ w = \mathrm{Tr}_\iota(e) = \mathrm{Tr}_{\bar{\iota}}(\bar{e}) = \dim(\iota_s) = \dim(\bar{\iota}_{\bar{s}}) = \sqrt{\dim(\theta_P)}$$

by Proposition 2.4, and

$P \circ X^* \circ (P \times P) \circ X \circ P =$ 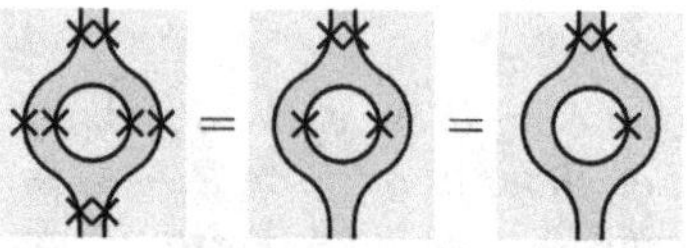 $= \dim(\iota_s) \cdot P,$

hence $n_P = x_P^* x_P = \dim(\iota_s) \cdot 1_{\theta_P} = \sqrt{\dim(\theta_P)} \cdot 1_{\theta_P}$. (Contact with Corollary 4.2 is made by noting that $\tilde{w}_P^* \tilde{w}_P = w_P^* n_P w_P = \sqrt{\dim(\theta_P)} \cdot w_P^* w_P = \dim(\theta_P)$.) □

**Corollary 4.10** *If $p$ or $\bar{p}$ satisfy the conditions in Proposition* 4.9, *the same is true for* $1 - p$ *resp.* $1 - \bar{p}$. *Thus, every simple Q-system with* $\dim \mathrm{Hom}(\mathrm{id}, \theta) > 1$ *has a decomposition into irreducible Q-systems* $\mathbf{A}_P$ *with* $\dim \mathrm{Hom}(\mathrm{id}, \theta_P) = 1$.

Namely, if $\mathbf{A}_P$ is reducible, one can just continue the decomposition.

(Notice, however, that unlike in Sect. 4.2 the decomposition corresponds to a partition of unity by $p_i$, not by $P_i = \bar{p}_i p_i$! This reflects the obvious fact that $[\theta] = (\bigoplus_n [\bar{\iota}_n])(\bigoplus_m [\iota_m])$ is different from $\bigoplus[\theta_P] \simeq \bigoplus[\bar{\iota}_n\iota_n]$.)

Finally, instead of characterizing the projection $p = \bar{\iota}(e) \in \mathrm{Hom}(\theta, \theta)$ satisfying the pair of relations as in Proposition 4.9, one may also write $e = \iota(q)v$ which is in $\mathrm{Hom}(\iota, \iota)$ iff $q \in \mathrm{Hom}(\theta, \mathrm{id})$, and characterize the operator $q$. Indeed, by Lemma 3.16, $e$ is idempotent iff $q = (q \times q) \circ x$, and $e$ is selfadjoint iff $q^* = (1_\theta \times q) \circ x \circ w$. In view of these properties, the first of the two conditions on $p = \theta(q)x$ is equivalent to $q^* = (q \times 1_\theta) \circ x \circ w$, whereas the second one is automatic. Therefore, $q \in \mathrm{Hom}(\theta, \mathrm{id})$ satisfying

$q \equiv$ 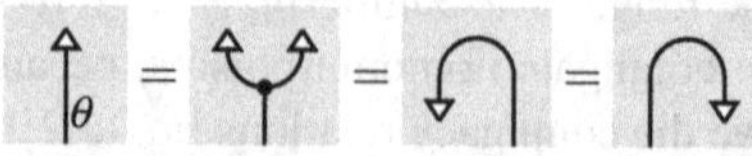 (4.3.13)

give rise to projections $e = \iota(q)v \in \mathrm{Hom}(\iota, \iota)$, hence $p = \bar{\iota}(e) = \theta(q)x \in \mathrm{Hom}(\theta, \theta)$, hence also $\bar{p} \in \mathrm{Hom}(\theta, \theta)$ as in the proposition, hence the sub-Q-system.

Notice that the last equality in Eq. (4.3.13) is an instance of Proposition 2.6, which applies since $M$ is a factor ($\mathbf{A}$ is simple).

## 4.4 Intermediate Q-Systems

In this section, we shall characterize decompositions of $\iota : N \to M$ as

$$\iota = \iota_2 \circ \iota_1$$

when $M$ is a factor, i.e., intermediate von Neumann algebras $\iota_1(N)$ between $N$ and $M$.

Let $N \subset L \subset M$ be an intermediate extension with $\iota = \iota_2 \circ \iota_1$, hence $\theta = \bar{\iota}_1\theta_2\iota_1$. Let $\mathbf{A} = (\theta, w, x)$ and $\mathbf{A}_2 = (\theta_2, w_2, x_2)$ be the Q-systems for $N \subset M$ and $N \subset L$, respectively. The projection $e_2 = d_2^{-1} \cdot w_2 w_2^* \in \mathrm{Hom}(\theta_2, \theta_2)$ onto $\mathrm{id}_L \prec \theta_2$ defines a projection $P = \bar{\iota}_1(e_2) =$  $\in \mathrm{Hom}(\theta, \theta)$. The projection $P$ satisfies the relations Eq. (4.1.1) and

$P \circ w = w :$ 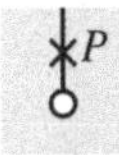 $= d_2^{-1} \cdot$ 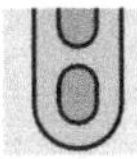 $=$ 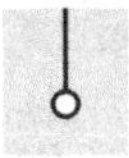 , (4.4.1)

hence $w^* \circ P \circ w = w^* \circ w = d \cdot \mathbf{1}_N$. It also satisfies

$x^* \circ (P \times P) \circ x = d_2^{-2} \cdot$ 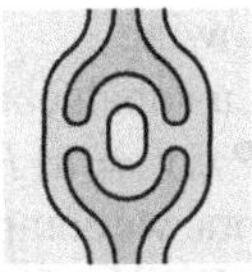 $= d_{\mathbf{A}} d_2^{-2} \cdot P.$

Conversely, the intermediate extension is characterized by the projection $P$:

**Proposition 4.11** *Let $\mathbf{A} = (\theta, w, x)$ be a Q-system in $\mathcal{C} \subset \mathrm{End}_0(N)$, defining $\iota : N \to M$ of dimension $\dim(\iota) = d_{\mathbf{A}}$. Let $P \in \mathrm{Hom}(\theta, \theta)$ be a projection satisfying Eqs. (4.1.1) and (4.4.1). Then Eq. (4.1.2) defines a reduced Q-system $\mathbf{A}_P$. The intermediate Q-system corresponds to the intermediate von Neumann algebra $N \subset L_P \subset M$ given by*

$$L_P := \iota(NP)v. \tag{4.4.2}$$

We will also refer to this reduced Q-system as **intermediate Q-system** of $\mathbf{A}$.

*Remark 4.12* A similar characterization of intermediate subfactors by projections has been given for the type II case in [2].

*Remark 4.13* The "normalization intertwiner" $n_P \in \mathrm{Hom}_0(\theta_P, \theta_P)$ as in Lemma 4.1 will in general not be a multiple of $1_{\theta_P}$, or equivalently, $x^* \circ (P \times P) \circ x$ will not be a multiple of $P$. Because of Corollary 3.6 and Lemma 3.16, this can only occur when $L_P$ is not a factor. We shall present an example below (Example 4.14). On the other hand, when $\dim \mathrm{Hom}(\mathrm{id}, \theta_P) = 1$, then $n_P \in \mathrm{Hom}_0(\theta_P, \theta_P)$ is trivially a multiple of $1_{\theta_P}$. In particular, when $N \subset M$ is irreducible, hence $\dim \mathrm{Hom}(\mathrm{id}, \theta) = 1$, then $N \subset L_P$ is irreducible, and $L_P$ is a factor. We also have: if $n_P \in \mathrm{Hom}_0(\theta_P, \theta_P) = \mu \cdot 1_{\theta_P}$, then $\mu = \dim(\theta_P)/d_{\mathbf{A}}$, because by Eq. (4.4.3), $r^*(P \times P)r = r^*(1_{\theta_P} \times P)r = \mathrm{Tr}_\theta(P) = \dim(\theta_P)$, while on the other hand, by Eq. (4.4.1), $r^*(P \times P)r = w^*x^*(P \times P)xw = \mu \cdot w^*Pw = \mu \cdot w^*w = \mu \cdot d_{\mathbf{A}}$.

*Proof of Proposition* 4.11 We first observe that by the assumed relations,

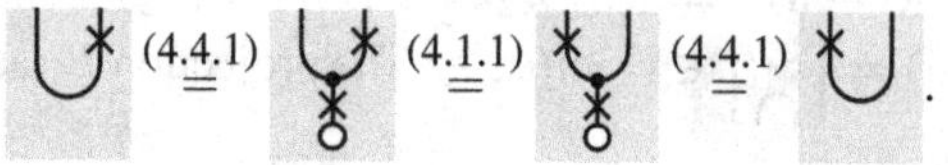


(4.4.3)

Thus, by Proposition 2.4,

$$r^* \circ (P \times P) \circ r = r^* \circ (1_\theta \times P) \circ r = \mathrm{Tr}_\theta(P) = \dim(\theta_P).$$

Hence, by Corollary 4.2, $\mathbf{A}_P = (\theta_P, w_P, x_P)$ is a reduced Q-system.

We write $\iota(n) \equiv n$ in the following.

To show that $L_P = NPv$ is a subalgebra of $M$, we compute $(n_1Pv)(n_2Pv) = n_1P\theta(n_2P)xv = n_1\theta(n_2)P\theta(P)xv = n_1P\theta(n_2)xPv$, using Eq. (4.1.1) in the last step. To show that $L_P$ is a *-algebra, we compute $(nPv)^* = r^*vPn^* = r^*\theta(Pn^*)v = r^*P\theta(n^*)v = r^*\theta(n^*)Pv$, using Eq. (4.4.3) in the third step. $L_P = N \cdot Pv$ is clearly weakly closed, and it is contained in $N \cdot v = M$.

We now compute the Q-system for $N \subset L_P$. Let $P = SS^*$ with $S \in N$, $S^*S = \mathbf{1}_N$, and put $\widetilde{w}_P := S^*w$ and $\widetilde{v}_P := S^*v \in L_P$. Then the embedding $\iota_P : N \to L_P$ is given by

$$\iota_P(n) \equiv n = nw^*v = nw^*Pv = nw^*SS^*v = n\widetilde{w}_P^*\widetilde{v}_P.$$

The conjugate map

$$\bar{\iota}_P(\cdot) := S^*\bar{\iota}(\cdot)S$$

is a homomorphism by Eq. (4.1.1), because every element of $L_P$ is of the form $nPv = nS\widetilde{v}_P$ with $n \in N$.

We claim that the pair $(\widetilde{w}_P, \widetilde{v}_P)$ solves the conjugacy relations Eq. (2.2.1) for $(\iota_P, \bar{\iota}_P)$. Certainly, $\widetilde{w}_P \in \mathrm{Hom}(\mathrm{id}_N, \bar{\iota}_P\iota_P)$, because

$$\bar{\iota}_P\iota_P(n) = S^*\bar{\iota}(n)S = S^*\theta(n)S = \theta_P(n).$$

Furthermore, $\widetilde{v}_P \in \mathrm{Hom}(\mathrm{id}_{L_P}, \iota_P\bar{\iota}_P)$ because $\widetilde{v}_P n = S^*vn = S^*\theta(n)v = S^*\theta(n)SS^*v = \theta_P(n)S^*v = \theta_P(n)\widetilde{v}_P$, and $\widetilde{v}_P\widetilde{v}_P = S^*vS^*v = S^*\theta(S^*)xv = S^*\theta(S^*)xSS^*v = \widetilde{x}_P\widetilde{v}_P = \bar{\iota}_P(\widetilde{v}_P)\widetilde{v}_P$, using Eq. (4.1.1) in the third step. The conjugacy relations then follow from Eq. (4.1.1).

Finally, $\bar{\iota}(\widetilde{v}_P) = S^*\bar{\iota}(S^*v)S = S^*\theta(S^*)xS = \widetilde{x}_P$. Thus, after the appropriate rescaling as in Corollary 4.2, the Q-system for $N \subset L_P$ coincides with the reduced Q-system $\mathbf{A}_P = (\theta_P, w_P, x_P)$. □

*Example 4.14* We give here a counterexample, showing that $n_P$ is *not necessarily* a multiple of $1_{\theta_P}$.

Let $N \subset L \subset M$, where $N$ and $M$ are factors, and $L = \bigoplus_i L_i$ a finite direct sum of factors. Let $\iota : N \to M$ given by $[\iota] = \bigoplus_i [\iota_{2i}\iota_{1i}]$ where $\iota_{1i} : N \to L_i$ and $\iota_{2i} : L_i \to M$. Similarly, $[\bar{\iota}] = \bigoplus_i [\bar{\iota}_{1i}\bar{\iota}_{2i}]$ where $\bar{\iota}_{1i} : L_i \to N$ and $\bar{\iota}_{2i} : M \to L_1$. We choose orthonormal isometries $s_i \in \mathrm{Hom}(\iota_{2i}\iota_{1i}, \iota)$ and $t_i \in \mathrm{Hom}(\bar{\iota}_{1i}\bar{\iota}_{2i}, \bar{\iota})$. The canonical endomorphism is $[\theta] = [\bar{\iota}\iota] = \bigoplus_{ij}[\bar{\iota}_{1i}\bar{\iota}_{2i}\iota_{2j}\iota_{1j}]$.

The intermediate embedding is described by $\iota_1 = \bigoplus_i \iota_{1i} : N \to L$, as in Sect. 2.3, with canonical endomorphism $[\theta_1] = \bigoplus_i [\bar{\iota}_{1i}\iota_{1i}] \prec [\theta]$.

For $N \subset M$ we construct a standard Q-system as usual (cf. Lemma 2.1): with standard pairs $(w_{1i}, \overline{w}_{1i})$ for $\iota_{1i}(N) \subset L_i$ and $(w_{2i}, \overline{w}_{2i})$ for $\iota_{2i}(L_i) \subset M$, we have the "composite" standard pairs as in Lemma 2.1(i)

$$(w_i = \bar{\iota}_{1i}(w_{2i})w_{1i}, \quad \overline{w}_i = \iota_{2i}(\overline{w}_{1i})\overline{w}_{2i})$$

for $\iota_i(N) = \iota_{2i}\iota_{1i}(N) \subset M$. Then $w \in \mathrm{Hom}(\mathrm{id}_N, \bar{\iota}\iota)$ and $\overline{w} \in \mathrm{Hom}(\mathrm{id}_M, \iota\bar{\iota})$ given by

$$w = \sum_i (t_i \times s_i) \circ w_i = \sum_i$$ 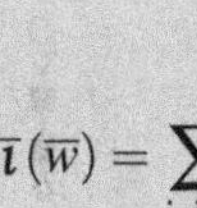  , $$\overline{w} = \sum_i (s_i \times t_i) \circ \overline{w}_i = \sum_i$$  

form a standard pair for $\iota : N \to M$, hence $(\theta, w, x)$ is the Q-system for $\iota(N) \subset M$, where

$$x = \bar{\iota}(\overline{w}) = \sum_{i,j,k}$$ 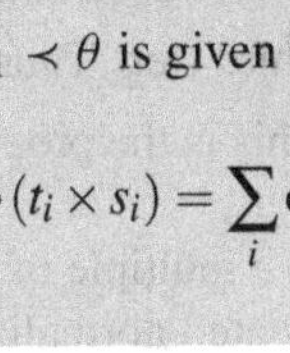  .

The projection $P \in \mathrm{Hom}(\theta, \theta)$ onto $\theta_1 \prec \theta$ is given by

$$P = \sum_i (t_i \times s_i) \circ (1_{\bar{\iota}_{1i}} \times E_{2i} \times 1_{\iota_{1i}}) \circ (t_i \times s_i) = \sum_i \dim(\iota_{2i})^{-1} \cdot$$  

where $E_{2i} = \dim(\iota_{2i})^{-1} \cdot w_{2i} w_{2i}^* \in \mathrm{Hom}(\bar\iota_{2i}\iota_{2i}, \bar\iota_{2i}\iota_{2i})$ is the projection onto $\mathrm{id}_{L_i} \prec \bar\iota_{2i}\iota_{2i}$. Then, one computes

$$x^* \circ (P \times P) \circ x = \sum_i \frac{\dim(\iota_{1i})}{\dim(\iota_{2i})} \cdot (t_i \times s_i) \circ (1_{\bar\iota_{1i}} \times E_{2i} \times 1_{\iota_{1i}}) \circ (t_i \times s_i).$$

Since $\frac{\dim(\iota_{1i})}{\dim(\iota_{2i})}$ in general depends on $i$, this is not a multiple of $P$ in general. In contrast, the normalization condition in [3] (cf. Remark 2.8) would be satisfied.

The following Lemma states how modules restrict to modules of intermediate Q-systems:

**Lemma 4.15** *If* $\mathbf{A}$ *is a Q-system, and* $\mathbf{A}_P$ *is an intermediate Q-system, then a (left)* $\mathbf{A}$*-module* $\mathbf{m} = (\beta, m)$ *restricts to a (left)* $\mathbf{A}_P$*-module*

$$\mathbf{m}_P = (\beta, m_P) \quad \textit{with} \quad m_P := \dim(\theta_P)^{\frac{1}{4}} \cdot (n_P^{-\frac{1}{2}} S^* \times 1_\beta) \circ m,$$

*where* $S^*S = 1$, $SS^* = P$. *If* $n_P \in \mathrm{Hom}_0(\theta_P, \theta_P)$ *is a multiple of* $1_{\theta_P}$, *then the normalization factor equals* $\dim(\theta_P)^{\frac{1}{4}} \cdot n_P^{-\frac{1}{2}} = (d_\mathbf{A}/d_{\mathbf{A}_P})^{\frac{1}{2}}$. *If* $\mathbf{m}$ *is standard, then* $\mathbf{m}_P$ *is standard. The analogous statements hold for right modules and bimodules.*

*Proof* One easily verifies, using Eq. (4.1.1), that the defining unit and representation properties of a module are satisfied. As for standardness of $\mathbf{m}_P$, one has

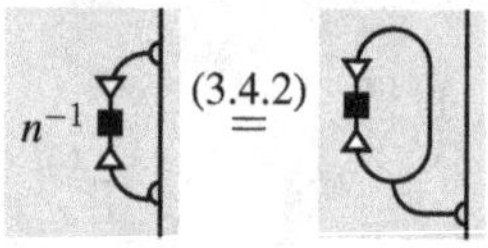

,

with

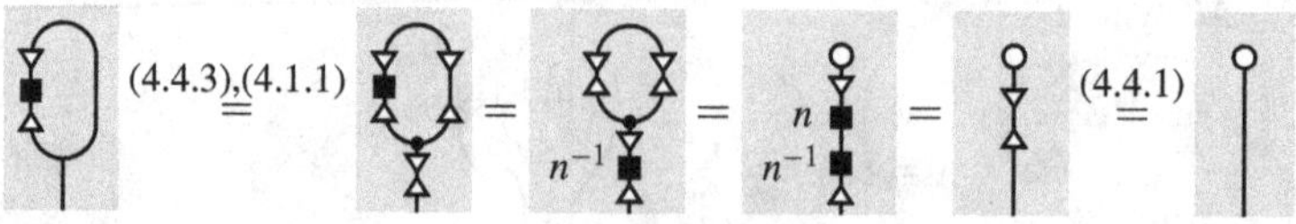

,

where in the second step, we have used that $n^{-1} \in \mathrm{Hom}_0(\theta_P, \theta_P)$, and the definition of $n_P$ and Eq. (4.4.1) in the third step. Thus, $m_P^* m_P = \dim(\theta_P)^{\frac{1}{2}} \cdot 1_\beta$ by Eq. (3.4.1). Because $\dim(\theta_P)^{\frac{1}{2}} = d_{\mathbf{A}_P}$, this is the proper normalization of a standard module in accord with Lemma 3.22.

If $n_P \in \mathrm{Hom}_0(\theta_P, \theta_P)$ is a multiple of $1_{\theta_P}$, then $n_P = \dim(\theta_P)/\dim(\theta)^{\frac{1}{2}} \cdot 1_{\theta_P}$ by Remark 4.13, giving the stated normalization factor.

The right module and bimodule cases are proven similarly. □

## 4.5 Q-Systems in Braided Tensor Categories

Let now $\mathscr{C} \subset \mathrm{End}_0(N)$ be in addition be *braided*. The braiding is denoted by

$$\varepsilon_{\rho,\sigma} \equiv$$ 

$$\in \mathrm{Hom}(\rho\sigma, \sigma\rho).$$

We also write $\varepsilon^{+}_{\rho,\sigma} \equiv \varepsilon_{\rho,\sigma}$, and $\varepsilon^{-}_{\rho,\sigma} \equiv \varepsilon^{*}_{\sigma,\rho}$ for the opposite braiding.

**Definition 4.16** If $\mathscr{C}$ is a braided C* tensor category with braiding $\varepsilon \equiv \varepsilon^{+}$, then $\mathscr{C}^{\mathrm{opp}}$ is the braided C* tensor category, which coincides with $\mathscr{C}$ as a C* tensor category, equipped with the opposite braiding $\varepsilon^{-}$.

*Remark 4.17* This definition is tantamount to the more fundamental definition (as in Sect. 3.3), according to which the monoidal product is regarded as a functor $\times : \mathscr{C} \times \mathscr{C} \to \mathscr{C}$, and $\mathscr{C}^{\mathrm{opp}}$ is the category equipped with the opposite monoidal product $\sigma \times^{\mathrm{opp}} \rho = \rho \times \sigma$. The braiding is a natural transformation between the functors $\times$ and $\times^{\mathrm{opp}}$, and its inverse : $\times^{\mathrm{opp}} \to \times$ is the opposite braiding. The equivalence can be seen "by left-right reflection of every diagram".

The cases of interest in QFT are $\mathscr{C} = \mathscr{C}^{\mathrm{DHR}}(\mathscr{A})$, the categories of DHR endomorphisms of local quantum field nets. These categories are braided categories, where the DHR braiding is defined in terms of unitary "charge transporters" changing the localization of DHR endomorphisms, as exposed Sect. 5.1.3. In low dimensions, the braiding and the opposite braiding arise, depending on the choice of a connected component of the spacelike complement. In particular, for a two-dimensional conformal net $\mathscr{A}_2 = \mathscr{A}_+ \otimes \mathscr{A}_-$ arising as a product of its two chiral subnets, we have $\mathscr{C}^{\mathrm{DHR}}(\mathscr{A}_2) = \mathscr{C}^{\mathrm{DHR}}(\mathscr{A}_+) \boxtimes \mathscr{C}^{\mathrm{DHR}}(\mathscr{A}_-)^{\mathrm{opp}}$, cf. Sect. 5.2.2.

**Definition 4.18** If $\rho \in \mathscr{C}$, then the operator

$$\mathrm{LTr}_{\rho}(\varepsilon_{\rho,\rho}) \equiv$$ 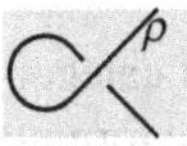

$$= \mathrm{RTr}_{\rho}(\varepsilon_{\rho,\rho}) \in \mathrm{Hom}(\rho,\rho)$$

is called the **twist**. The twist is a unitary self-intertwiner [4, afterLemma 4.3], [5, Proposition 2.4]; in particular, it is a complex phase denoted $\kappa_\rho$ if $\rho$ is irreducible.

*Example 4.19* (Braiding of the Ising category) The tensor category Example 3.1 can be equipped with four inequivalent braidings.

The braiding of the DHR category of the Ising model is specified by

$$\varepsilon_{\tau,\tau} = -1, \quad \varepsilon_{\sigma,\sigma} = \kappa_{\sigma}^{-1} \cdot rr^* + \kappa_{\sigma}^{3} \cdot tt^*, \quad \varepsilon_{\sigma,\tau} = \varepsilon_{\tau,\sigma} = -iu,$$

where $\kappa_\sigma = \exp \frac{2\pi i}{16}$.

(This braiding and its opposite, and a second pair of braidings obtained by replacing $\kappa_\sigma$ by $-\kappa_\sigma$, exhaust all possibilities. The second tensor category mentioned in Example 3.1 also admits four inequivalent braidings.)

**Definition 4.20** A Q-system $(\theta, w, x)$ in a braided tensor category is called **commutative** if

$$\varepsilon_{\theta,\theta} \circ x = x :$$
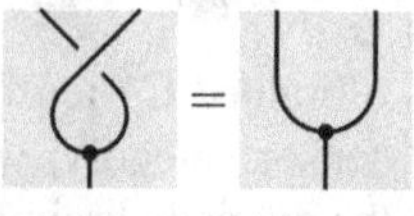
(4.5.1)

**Proposition 4.21** ([6]) *The canonical Q-system (cf. Proposition* 3.19) *of a braided C* tensor category is a commutative Q-system in the Deligne product* $\mathscr{C} \boxtimes \mathscr{C}^{\mathrm{opp}}$.

In local quantum field theory, commutative Q-systems describe local extensions of a given local quantum field theory [6], cf. Sect. 5.2.1.

Recall that the DHR category of a two-dimensional QFT which is the tensor product $\mathscr{A}_2 = \mathscr{A}_+ \otimes \mathscr{A}_-$ of two chiral QFTs, is $\mathscr{C}^{\mathrm{DHR}}(\mathscr{A}_2) = \mathscr{C}^{\mathrm{DHR}}(\mathscr{A}_+) \boxtimes \mathscr{C}^{\mathrm{DHR}}(\mathscr{A}_-)^{\mathrm{opp}}$ as a braided category. Therefore, if $\mathscr{A}_+$ and $\mathscr{A}_-$ are isomorphic, the 2D extension associated with the canonical Q-system in $\mathscr{C}^{\mathrm{DHR}}(\mathscr{A}) \boxtimes \mathscr{C}^{\mathrm{DHR}}(\mathscr{A})^{\mathrm{opp}}$ is always a local QFT.

## 4.6 $\alpha$-Induction

If $\mathbf{A} = (\theta, w, x)$ is a Q-system in a braided category, then $\mathbf{m} = (\beta = \theta\rho, m = \theta^2(\varepsilon^{\pm}_{\theta,\rho})x^{(2)})$ is a standard **A**-**A**-bimodule. The formula Eq. (3.6.3) for the associated endomorphism $\varphi : M \to M$ becomes

$$\varphi(\iota(n)v) = \iota(\rho(n)\varepsilon^{\pm}_{\theta,\rho})v,$$

which is known as the $\alpha$-induction of $\rho \in \mathrm{End}_0(N)$ to $\alpha^{\mp}_\rho \in \mathrm{End}_0(M)$, originally defined by $\bar\iota \circ \alpha^{\pm}_\rho = \mathrm{Ad}_{\varepsilon_{\rho,\theta}} \circ \rho \circ \bar\iota$ [6–9].

The endomorphisms $\alpha^{\pm}_\rho$ *extend* the endomorphism $\rho \in \mathrm{End}(N)$:

$$\alpha^{\pm}_\rho \circ \iota = \iota \circ \rho, \tag{4.6.1}$$

and the mappings $\rho \mapsto \alpha^{\pm}_\rho, t \to \iota(t)$ are functorial, namely if $t \in \mathrm{Hom}(\rho_1, \rho_2)$, then

$$\iota(t) \in \mathrm{Hom}(\alpha^{\pm}_{\rho_1}, \alpha^{\pm}_{\rho_2}). \tag{4.6.2}$$

However, $\iota : \mathrm{Hom}(\rho_1, \rho_2) \to \mathrm{Hom}(\alpha^{\pm}_{\rho_1}, \alpha^{\pm}_{\rho_2})$ is in general not surjective. E.g., $\alpha^{\pm}_{\rho}$ may possess self-intertwiners (i.e., $\alpha^{\pm}_{\rho}$ is reducible), while $\rho$ is irreducible.

**Corollary 4.22** (i) *One has* $\alpha^{\pm}_{\overline{\rho}} = \overline{\alpha}^{\pm}_{\rho}$ *and* $\dim(\alpha^{\pm}_{\rho}) = \dim(\rho)$.
(ii) *If* $(\theta, w, x)$ *is a Q-system in* $\mathrm{End}_0(N)$*, then* $(\alpha^{\pm}_{\theta}, \iota(w), \iota(x))$ *is a Q-system in* $\mathrm{End}_0(M)$.

*Proof* Since conjugacy and dimension are defined in terms of intertwiners and their algebraic relations, (i) follows from Eq. (4.6.2). Similarly, (ii) follows because also Q-systems are defined in terms of intertwiners and their algebraic relations. □

If the category $\mathscr{C}$ is modular (cf. Sect. 4.11), then the matrices

$$Z_{\rho,\sigma} = \dim\mathrm{Hom}(\alpha^{-}_{\rho}, \alpha^{+}_{\sigma}) \tag{4.6.3}$$

are "modular invariants", i.e., they commute with the unitary representation of the modular group $SL(2, \mathbb{Z})$ defined by the braiding [10–12], and have many other remarkable properties [11–13] that can, not least, be exploited for classifications and actual computations.

## 4.7 Mirror Q-Systems

Let $N \otimes \widetilde{N} \subset M$ be an irreducible finite-index subfactor, and $\mathbf{A} = (\Theta, W, X)$ its Q-system. The subfactor is called a canonical tensor product subfactor (CTPS), if $\Theta$ has the form

$$[\Theta] = \bigoplus Z_{\rho,\widetilde{\sigma}}[\rho] \otimes [\widetilde{\sigma}],$$

where $\rho \in \mathrm{End}_0(N)$ and $\widetilde{\sigma} \in \mathrm{End}_0(\widetilde{N})$ are irreducible, and $Z_{\rho,\widetilde{\sigma}}$ are multiplicities.

The following proposition was derived in [10, Theorem 3.6]:

**Proposition 4.23** *The following are equivalent:*

(i) $[\mathrm{id}] \otimes [\widetilde{\sigma}] \prec \Theta$ *implies* $[\widetilde{\sigma}] = [\mathrm{id}_{\widetilde{N}}]$*, and* $[\sigma] \otimes [\mathrm{id}] \prec \Theta$ *implies* $[\sigma] = [\mathrm{id}_N]$.
(ii) *It holds*

$$(N \otimes \mathbf{1})' \cap M = (\mathbf{1} \otimes \widetilde{N}), \qquad (\mathbf{1} \otimes \widetilde{N})' \cap M = (N \otimes \mathbf{1}).$$

(iii) *There is a bijection* $F$ *between the set* $\Delta$ *of sectors* $[\rho]$ *and the set* $\widetilde{\Delta}$ *of sectors* $[\widetilde{\sigma}]$ *contributing to* $\Theta$ *such that*

$$[\Theta] = \bigoplus [\rho] \otimes F[\rho];$$

*$\Delta$ and $\widetilde{\Delta}$ are closed under fusion (i.e., the product $[\rho_1][\rho_2]$ decomposes into irreducibles in $\Delta$ resp. $\widetilde{\Delta}$), and $F$ is an isomorphism of fusion rings.*

Under stronger conditions (the tensor categories generated by the endomorphisms $\rho \in [\rho] \in \Delta$ and $\widetilde{\rho} \in [\widetilde{\rho}] \in \widetilde{\Delta}$ are braided and modular, and $\mathbf{A}$ is commutative), the isomorphism of fusion rings is even an isomorphism of braided tensor categories [14].

The canonical Q-systems in $\mathcal{C} \boxtimes \mathcal{C}^{\rm opp}$ with $[\Theta] = \bigoplus [\rho] \otimes [\overline{\rho}]$ and $F[\rho] = [\overline{\rho}]$, cf. Corollary 3.21, are examples fulfilling the properties in Proposition 4.23.

Xu [15] has strengthened the statement:

**Proposition 4.24** *Assume that the equivalent conditions of Proposition 4.23 are fulfilled. Let $\mathcal{C} \subset \mathrm{End}_0(N)$ be the full tensor subcategory generated by endomorphisms $\rho \in [\rho] \in \Delta$, and similarly $\widetilde{\mathcal{C}} \subset \mathrm{End}_0(\widetilde{N})$. If $\mathcal{C}$ and $\widetilde{\mathcal{C}}$ are braided categories, hence $\mathbf{A}$ is a Q-system in the braided category $\mathcal{C} \boxtimes \widetilde{\mathcal{C}}$, the $\alpha$-induction of $\rho \otimes \widetilde{\sigma} \in \mathcal{C} \otimes \widetilde{\mathcal{C}}$ is well-defined (choosing $\alpha^+$ for definiteness). Then, one has*

$$\iota(\mathrm{Hom}(\rho_1, \rho_2) \otimes \mathbf{1}) = \mathrm{Hom}(\alpha_{\rho_1 \otimes \mathrm{id}}, \alpha_{\rho_2 \otimes \mathrm{id}}) \tag{4.7.1}$$

*(and similar for $\widetilde{\rho}$), rather than just the inclusion $\subset$ according to Eq. (4.6.2). Moreover, if $[\widetilde{\rho}] = F[\rho]$, then $\alpha_{\mathrm{id}\otimes\widetilde{\rho}}$ and $\alpha_{\overline{\rho}\otimes\mathrm{id}}$ are unitarily equivalent. If $\mathbf{A}$ is commutative, the unitary $u \in \mathrm{Hom}(\alpha_{\rho\otimes\mathrm{id}}, \alpha_{\mathrm{id}\otimes\widetilde{\rho}})$ can be chosen such that*

$$(u \times u) \circ \iota(\varepsilon_{\rho,\rho} \otimes \mathbf{1}) = \iota(\mathbf{1} \otimes \varepsilon^*_{\widetilde{\rho},\widetilde{\rho}}) \circ (u \times u). \tag{4.7.2}$$

From this, he concludes the existence of the "mirror extension" defined by a "mirror Q-system" in $\widetilde{\mathcal{C}}$ associated with a Q-system in $\mathcal{C}$, as follows.

Assume that the equivalent conditions of Proposition 4.23 are fulfilled. If $(\theta, w, x)$ is a Q-system in $\mathcal{C}$, there is $\widetilde{\theta}$ such that $\alpha_{\mathrm{id}\otimes\widetilde{\theta}}$ and $\alpha_{\theta\otimes\mathrm{id}}$ are unitarily equivalent, i.e., $[\widetilde{\theta}] = F[\theta]$. Let $u \in \mathrm{Hom}(\alpha_{\theta\otimes\mathrm{id}}, \alpha_{\mathrm{id}\otimes\widetilde{\theta}})$ unitary. Then, by Eq. (4.7.1),

$$u \circ \iota(w \otimes \mathbf{1}) \in u \circ \mathrm{Hom}(\mathrm{id}, \alpha_{\theta\otimes\mathrm{id}}) = \mathrm{Hom}(\mathrm{id}, \alpha_{\mathrm{id}\otimes\widetilde{\theta}}) = \iota(\mathbf{1} \otimes \mathrm{Hom}(\mathrm{id}, \widetilde{\theta})),$$

and similarly

$$(u \times u) \circ \iota(x \otimes \mathbf{1}) \circ u^* \in \iota(\mathbf{1} \otimes \mathrm{Hom}(\widetilde{\theta}, \widetilde{\theta}^2)).$$

This defines $\widetilde{w}$ and $\widetilde{x}$ such that $u \circ \iota(w \otimes \mathbf{1}) = \iota(\mathbf{1} \otimes \widetilde{w})$ and $(u \times u) \circ \iota(x \otimes \mathbf{1}) \circ u^* = \iota(\mathbf{1} \otimes \widetilde{x})$.

**Corollary 4.25** ([15, Theorem 3.8]) *$(\widetilde{\theta}, \widetilde{w}, \widetilde{x})$ is a Q-system in $\widetilde{\mathcal{C}}$. If $\mathbf{A} = (\Theta, W, X)$ is commutative, then $(\widetilde{\theta}, \widetilde{w}, \widetilde{x})$ is commutative iff $(\theta, w, x)$ is commutative.*

*Proof* The defining relations for $(\widetilde{\theta}, \widetilde{w}, \widetilde{x})$ to be a Q-system are satisfied because by Corollary 4.22 $(\alpha_{\theta\otimes\mathrm{id}}, \iota(w \otimes \mathbf{1}), \iota(x \otimes \mathbf{1}))$ is a Q-system in $\mathrm{End}_0(M)$, and hence $(\alpha_{\mathrm{id}\otimes\widetilde{\theta}}, u \circ \iota(w \otimes \mathbf{1}), (u \times u) \circ \iota(x \otimes \mathbf{1}) \circ u^*)$ is an equivalent Q-system in $\mathrm{End}_0(M)$. If $\mathbf{A}$ is commutative, then Eq. (4.7.2) proves the second statement. □

## 4.8 Centre of Q-Systems

Let $\mathbf{A} = (\theta, w, x)$ be a Q-system of dimension $d_{\mathbf{A}}$ in a braided C* category $\mathscr{C}$, $r = x \circ w$, and $\mathbf{m} = (\beta, m)$ an **A**-**A**-bimodule. Define $Q^{\pm}_{\mathbf{m}} \in \mathrm{Hom}(\beta, \beta)$ by

$$Q^{\pm}_{\mathbf{m}} := (r^* \times 1_\beta) \circ (1_\theta \times \varepsilon^{\pm}_{\beta,\theta}) \circ m = (1_\beta \times r^*) \circ (\varepsilon^{\mp}_{\theta,\beta} \times 1_\theta) \circ m : \quad Q^{+}_{\mathbf{m}} =$$ 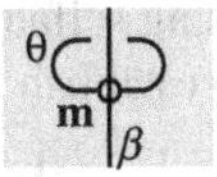

.

**Lemma 4.26** (cf. [16]) $P^{\pm}_{\mathbf{m}} := d_{\mathbf{A}}^{-1} \cdot Q^{\pm}_{\mathbf{m}}$ *are projections. For* $\mathbf{m} = \mathbf{A}$ *the trivial* **A**-**A**-*bimodule, the projections* $P^{\pm} \equiv P^{\pm}_{\mathbf{A}}$ *satisfy the relations*

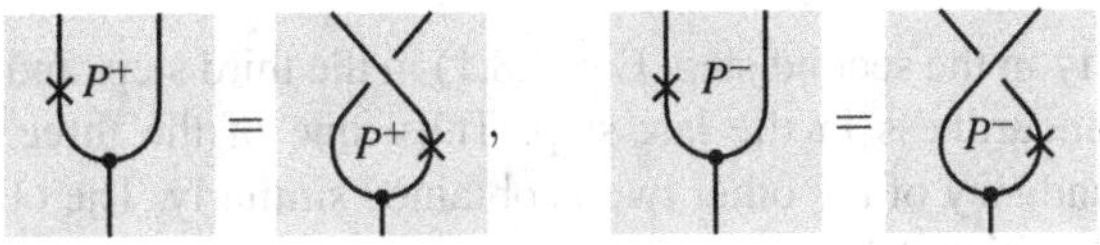


(4.8.1)

*Proof* We prove idempotency and selfadjointness of $P^{+}_{\mathbf{m}}$, using the representation property of the bimodule, the associativity of the Q-system, and the unitarity of the twist (cf. Definition 4.18) in the last step:

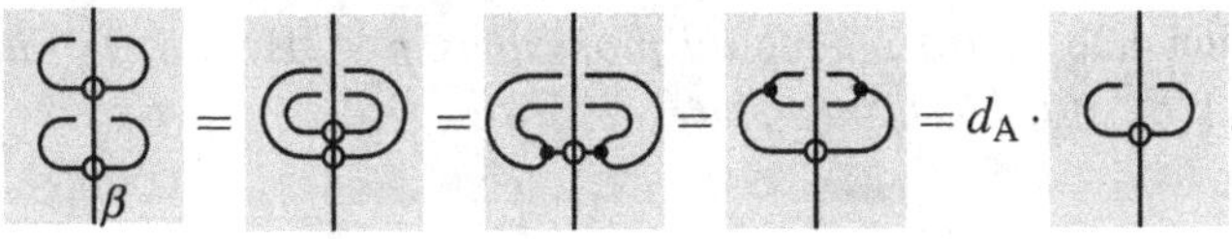


and

We then prove the relation for $P^+ \equiv P^+_{\mathbf{A}}$:

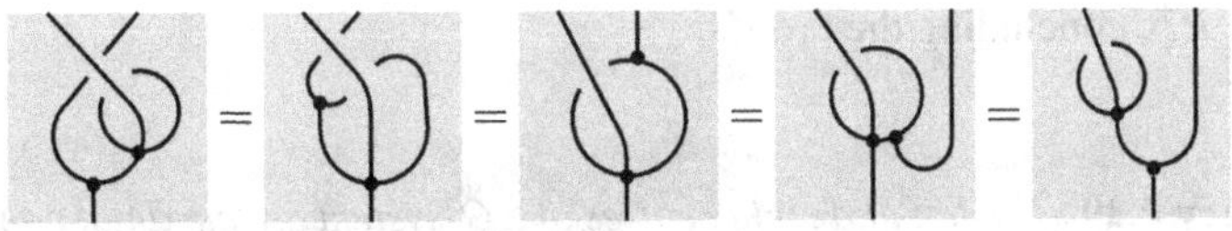

where we have several times used associativity of the Q-system. The proofs for $P^-$ are similar. □

**Lemma 4.27** (cf. [16]) *The projections* $P^{\pm}_{\mathbf{A}}$ *satisfy Eqs.* (4.1.1) *and* (4.4.1). *Hence, they define intermediate extensions by Proposition* 4.11*; the corresponding reduced Q-systems* $(\theta^{\pm}_{P}, w^{\pm}_{P}, x^{\pm}_{P})$ *are called* **left resp. right centre** $C^{\pm}[\mathbf{A}]$.

*Proof* We prove Eq. (4.4.1) by

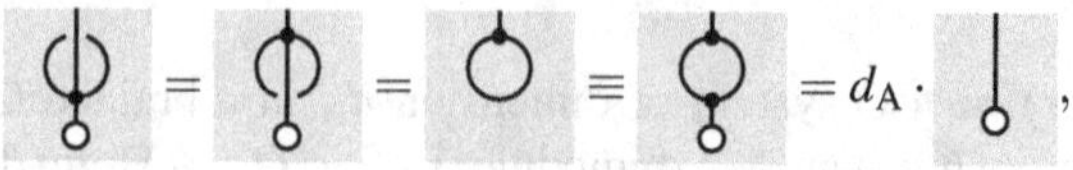


using selfadjointness of $P^{\pm}$, and the unit property and standardness of **A**. In order to establish Eq. (4.1.1) (for $P^+_{\mathbf{A}}$), we compute

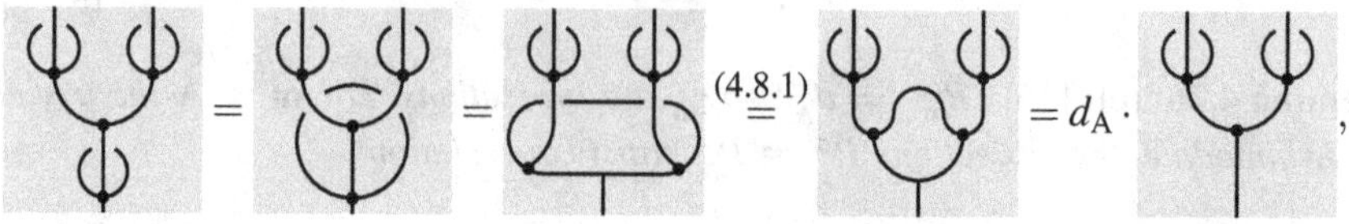


using associativity in the second step, Eq. (4.8.1) in the third step, and the Frobenius property and standardness in the last step. Thus, one of the three projections is redundant. Redundancy of the other two is obtained similarly. The other statements follow from Proposition 4.11. □

The left and right centre projections can be characterized as the maximal ones satisfying Eq. (4.8.1):

**Proposition 4.28** ([16]) *Among all projections $p \in \mathrm{Hom}(\theta, \theta)$ satisfying Eq. (4.8.1), $P^{\pm}_{\mathbf{A}}$ are the maximal ones.*

*Proof* For $P^+_{\mathbf{A}}$:

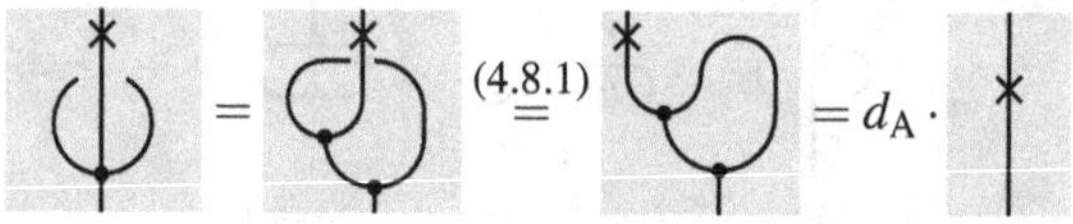


Thus, $p < P^+_{\mathbf{A}}$, concluding the proof. □

**Corollary 4.29** *The left and right centres of a Q-system are maximal commutative intermediate Q-systems. A Q-system **A** is commutative iff $P^+_{\mathbf{A}} = 1_\theta$ iff $P^+_{\mathbf{A}} = 1_\theta$ (i.e., $C^{\pm}[\mathbf{A}] = \mathbf{A}$).*

*Proof* Follows from Propositions 4.28 and 4.11 because by definition, a Q-system is commutative iff $1_\theta$ satisfies Eq. (4.8.1). □

This result is of interest in the applications to local QFT, where the intermediate extension associated with the centre projections can be identified as certain relative commutants of local algebras [1], cf. Sect. 5.2.3.

## 4.9 Braided Product of Q-Systems

**Definition 4.30** Let $\mathbf{A} = (\theta^{\mathbf{A}}, w^{\mathbf{A}}, x^{\mathbf{A}})$ and $\mathbf{B} = (\theta^{\mathbf{B}}, w^{\mathbf{B}}, x^{\mathbf{B}})$ be two Q-systems in a braided C* tensor category $\mathscr{C}$. Then there are two natural product Q-systems, called **braided products** and denoted as $\mathbf{A} \times^{\pm} \mathbf{B}$, given by the object $\theta = \theta^{\mathbf{A}}\theta^{\mathbf{B}}$ and the interwiners

$w = w^{\mathrm{A}} \times w^{\mathrm{B}} \equiv$ 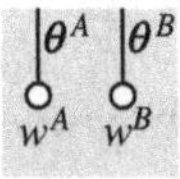

, $\quad x^{\pm} = (1_{\theta^{\mathrm{A}}} \times \varepsilon^{\pm}_{\theta^{\mathrm{A}},\theta^{\mathrm{B}}} \times 1_{\theta^{\mathrm{B}}}) \circ (x^{\mathrm{A}} \times x^{\mathrm{B}})$ : $x^{+} =$ 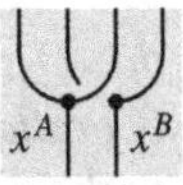

.

The extension $N \subset M^{\pm}$ corresponding to the braided product of two Q-systems is called the **braided product of extensions**.

Notice that $\dim \mathrm{Hom}(\mathrm{id}_N, \theta^{\mathbf{A}}\theta^{\mathbf{B}}) = \dim \mathrm{Hom}(\theta^{\mathbf{A}}, \theta^{\mathbf{B}})$ can in general be larger than 1, even if $\dim \mathrm{Hom}(\mathrm{id}_N, \theta^{\mathbf{A}}) = \dim \mathrm{Hom}(\mathrm{id}_N, \theta^{\mathbf{B}}) = 1$. Thus, the braided product of extensions is in general not irreducible, and not even a factor, even if both extensions are irreducible. We shall return to this issue below.

One can easily see that the braided product $\mathbf{A} \times^{\pm} \mathbf{B}$ contains both $\mathbf{A}$ and $\mathbf{B}$ as intermediate Q-systems, via the natural projections $d_{\mathbf{A}}^{-1} \cdot (w^{\mathbf{A}}w^{\mathbf{A}*} \times 1_{\theta^{\mathbf{B}}})$ onto $\theta^{\mathbf{A}} \prec \theta^{\mathbf{A}}\theta^{\mathbf{B}}$ and $d_{\mathbf{B}}^{-1} \cdot (1_{\theta^{\mathbf{A}}} \times w^{\mathbf{B}}w^{\mathbf{B}*})$ onto $\theta^{\mathbf{B}} \prec \theta^{\mathbf{A}}\theta^{\mathbf{B}}$, respectively.

Expressed in terms of the corresponding extensions, the braided products $N \subset M^{\pm}$ of extensions $N \subset M^{\mathbf{A}}$, $N \subset M^{\mathbf{B}}$ contain both $M^{\mathbf{A}}$ and $M^{\mathbf{B}}$ as intermediate extensions:

$$N \begin{array}{c} \subset M^{\mathbf{A}} \subset \\ \subset M^{\mathbf{B}} \subset \end{array} M^{\pm} \tag{4.9.1}$$

More precisely, we have

**Lemma 4.31** *The braided products $N \subset M^{\pm}$ of two extensions $N \subset M^{\mathbf{A}} = \iota^{\mathbf{A}}(N)v^{\mathbf{A}}$, $N \subset M^{\mathbf{B}} = \iota^{\mathbf{B}}(N)v^{\mathbf{B}}$ are generated by the subalgebra $N$ and the generator $v^{\pm} = v^{\mathbf{A}}v^{\mathbf{B}}$, where $v^{\mathbf{A}}$ and $v^{\mathbf{B}}$ are embedded into $M^{\pm}$ as*

$v^{\mathbf{A}} = \iota^{\pm}(\theta^{\mathbf{A}}(w^{\mathbf{B}*}))v^{\pm} =$ 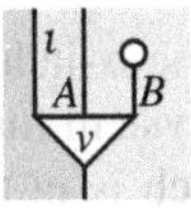

, $\quad v^{\mathbf{B}} = \iota^{\pm}(w^{\mathbf{A}*})v^{\pm} =$ 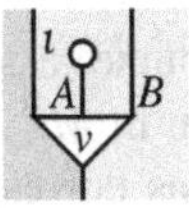

.

*Thus $M^{\pm}$ contain both $M^{\mathbf{A}} = \iota^{\pm}(N)v^{\mathbf{A}}$ and $M^{\mathbf{B}} = \iota^{\pm}(N)v^{\mathbf{B}}$ as intermediate algebras. In $M^{\pm}$, the generators $v^{\mathbf{A}}$ and $v^{\mathbf{B}}$ satisfy the relations*

$$v^{\mathbf{B}}v^{\mathbf{A}} = \iota(\varepsilon^{\pm}_{\theta^{\mathbf{A}},\theta^{\mathbf{B}}}) \cdot v^{\mathbf{A}}v^{\mathbf{B}}.$$

We can relate the braided product of Q-systems with the $\alpha$-induction of Q-systems, Corollary 4.22, as follows.

**Proposition 4.32** *Let* $\iota^{\mathbf{A}} : N \to M^{\mathbf{A}}$ *and* $\iota^{\mathbf{B}} : N \to M^{\mathbf{B}}$, *and* $\mathbf{A} = (\theta^{\mathbf{A}}, w^{\mathbf{A}}, x^{\mathbf{A}})$ *and* $\mathbf{B} = (\theta^{\mathbf{B}}, w^{\mathbf{B}}, x^{\mathbf{B}})$ *the associated Q-systems in a braided* $C^*$ *tensor category* $\mathscr{C} \subset \mathrm{End}_0(N)$. *Denote by*

$$\alpha^{\pm}(\mathbf{B}) = (\alpha^{\pm}_{\theta^{\mathbf{B}}}, \iota^{\mathbf{A}}(w^{\mathbf{B}}), \iota^{\mathbf{A}}(x^{\mathbf{B}}))$$

*the Q-system in* $\mathrm{End}_0(M^{\mathbf{A}})$ *obtained from* $\mathbf{B}$ *by* $\alpha$*-induction along* $\mathbf{A}$ *(Corollary 4.22(ii)). Then* $\alpha^{\pm}(\mathbf{B})$ *is the Q-system for the extension* $M^{\mathbf{A}} \subset M^{\mp}$ *in the diagram Eq. (4.9.1).*

*More precisely, if we write the extensions corresponding to the braided products* $\mathbf{A} \times^{\pm} \mathbf{B}$ *as* $\iota^{\pm} : N \to M^{\pm}$, *and the extension corresponding to* $\alpha^{\pm}(\mathbf{B})$ *as* $\jmath^{\mathbf{B}\pm} : M^{\mathbf{A}} \to M^{\alpha\pm}$, *such that* $\alpha^{\pm}_{\theta^{\mathbf{B}}} = \bar{\jmath}^{\mathbf{B}\pm}\jmath^{\mathbf{B}\pm}$, *then we have* $M^{\alpha\pm} = M^{\mp}$ *and*

$$\iota^{\mp} = \jmath^{\mathbf{B}\pm} \circ \iota^{\mathbf{A}}.$$

*Proof* It suffices to verify that the composite Q-system according to Lemma 2.1(i) arising by the composition of embeddings $\iota^{\mathbf{A}} : N \to M^{\mathbf{A}}$ and $\jmath^{\mathbf{B}\pm} : M^{\mathbf{A}} \to M^{\alpha\pm}$, coincides with $\mathbf{A} \times^{\mp} \mathbf{B} = (\Theta, W, X^{\mp})$. Indeed, by the definitions and Eq. (4.6.1) we have

$$\bar{\iota}^{\mathbf{A}}\bar{\jmath}^{\mathbf{B}\pm} \circ \jmath^{\mathbf{B}\pm}\iota^{\mathbf{A}} = \bar{\iota}^{\mathbf{A}}\alpha^{\pm}_{\theta^{\mathbf{B}}}\iota^{\mathbf{A}} = \bar{\iota}^{\mathbf{A}}\iota^{\mathbf{A}}\theta^{\mathbf{B}} = \theta^{\mathbf{A}}\theta^{\mathbf{B}} = \Theta,$$

$$\bar{\iota}^{\mathbf{A}}(\iota^{\mathbf{A}}(w^{\mathbf{B}}))w^{\mathbf{A}} = \theta_1(w^{\mathbf{B}})w^{\mathbf{A}} = W,$$

and, denoting the generator of $\alpha^{\pm}(\mathbf{B})$ by $v^{\pm}$, such that $\bar{\jmath}^{\mathbf{B}\pm}(v^{\pm}) = \iota^{\mathbf{A}}(x^{\mathbf{B}})$:

$$\begin{aligned}\bar{\iota}^{\mathbf{A}}\bar{\jmath}^{\mathbf{B}\pm}\big[\jmath^{\mathbf{B}\pm}(v^{\mathbf{A}})v^{\pm}\big] &= \bar{\iota}^{\mathbf{A}}\big[\alpha^{\pm}_{\theta^{\mathbf{B}}}(v^{\mathbf{A}})\iota^{\mathbf{A}}(x^{\mathbf{B}})\big] = \bar{\iota}^{\mathbf{A}}\big[\iota^{\mathbf{A}}(\varepsilon^{\mp}_{\theta^{\mathbf{A}},\theta^{\mathbf{B}}})v^{\mathbf{A}}\iota^{\mathbf{A}}(x^{\mathbf{B}})\big]\\ &= \theta^{\mathbf{A}}(\varepsilon^{\mp}_{\theta^{\mathbf{A}},\theta^{\mathbf{B}}})x^{\mathbf{A}}\theta^{\mathbf{A}}(x^{\mathbf{B}}) = X^{\mp}.\end{aligned}$$

□

Of course, a similar result is true for the $\alpha$-induction of the Q-system $\mathbf{A}$ to a Q-system in $M^{\mathbf{B}}$, namely $\alpha^{\pm}(\mathbf{A})$ is the Q-system for $M^{\mathbf{B}}$ in the braided product of extensions corresponding to $\mathbf{B} \times^{\mp} \mathbf{A}$, which is in turn unitarily equivalent to the braided product of extensions corresponding to $\mathbf{A} \times^{\pm} \mathbf{B}$.

As mentioned before, the braided product of two extensions may fail to be irreducible, or to be a factor, even if both extensions are irreducible. For the braided product of two *commutative* extensions, the centre equals the relative commutant. This result is of particular interest in the applications to local QFT, where phase boundaries are described by the braided product of two local extensions [1].

**Proposition 4.33** *Let* $\mathbf{A} = (\theta^{\mathbf{A}}, w^{\mathbf{A}}, x^{\mathbf{A}})$ *and* $\mathbf{B} = (\theta^{\mathbf{B}}, w^{\mathbf{B}}, x^{\mathbf{B}})$ *be two commutative Q-systems in a braided category, and* $\mathbf{A} \times^{\pm} \mathbf{B} = (\theta, w, x)$ *the product Q-system (with either braiding). Let* $N \subset M$ *be the corresponding braided product of extensions. Then the centre* $M' \cap M$ *of* $M$ *equals the relative commutant* $\iota(N)' \cap M$.

*Proof* In view of Lemma 3.16, we have to show that every $q \in \mathrm{Hom}(\theta^{\mathbf{A}}\theta^{\mathbf{B}}, \mathrm{id}_N)$ satisfies Eq. (3.2.4). Let $q \in \mathrm{Hom}(\theta^{\mathbf{A}}\theta^{\mathbf{B}}, \mathrm{id}_N)$. Then

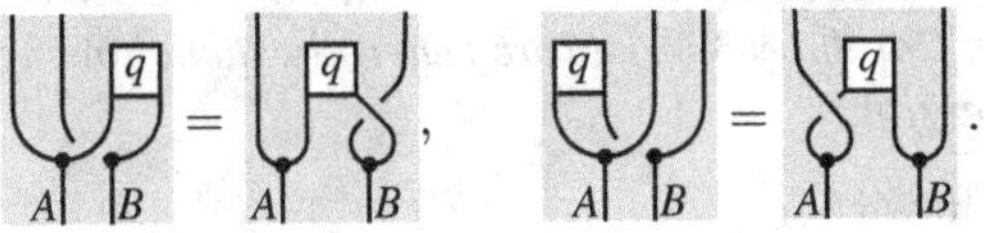


If both Q-systems are commutative, the two expressions are the same. □

## 4.10 The Full Centre

**Definition 4.34** ([16])Let $\mathbf{A} = (\theta, w, x)$ be a Q-system in $\mathscr{C}$. It trivially gives rise to a Q-system $\mathbf{A}\otimes\mathbf{1} = (\theta_i \otimes \mathrm{id}_N, w \otimes \mathbf{1}_N, x \otimes \mathbf{1}_N)$ in $\mathscr{C} \boxtimes \mathscr{C}^{\mathrm{opp}}$. Let $\mathbf{R}$ be the canonical Q-system in $\mathscr{C} \boxtimes \mathscr{C}^{\mathrm{opp}}$. Then the **full centre** of $\mathbf{A}$ is defined as the commutative Q-system in $\mathscr{C} \boxtimes \mathscr{C}^{\mathrm{opp}}$ given by the left centre of the $\times^{+}$-product

$$Z[\mathbf{A}] = C^{+}[(\mathbf{A} \otimes \mathbf{1}) \times^{+} \mathbf{R}]. \tag{4.10.1}$$

Because $\dim\mathrm{Hom}(\mathrm{id}, (\theta \otimes \mathrm{id})\Theta_{\mathrm{can}}) = \dim\mathrm{Hom}(\mathrm{id}, \theta)$, and the centre projection can only decrease multiplicities, the full centre is irreducible if $\mathbf{A}$ is irreducible. For a stronger statement, see Proposition 4.37.

**Proposition 4.35** ([17, Proposition 4.18]) *The full centre equals the* $\alpha$*-induction construction in* [10].

This result was conjectured in [18], and proven in [17]. In fact, it is rather easy to show that both the $\alpha$-induction construction and the full centre give

$$[\Theta] = \bigoplus Z_{\rho,\sigma}\, [\rho] \otimes [\overline{\sigma}]$$

with the multiplicities $Z_{\rho,\sigma}$ given by Eq. (4.6.3); whereas the equality of the respective intertwiners $X$ is more difficult to establish.

*Remark 4.36* The $\alpha$-induction construction [10] was originally found as a construction of two-dimensional local conformal QFT models out of chiral data, cf. Sect. 5.2.4. It is in fact a construction of commutative Q-systems in $\mathscr{C} \boxtimes \mathscr{C}^{\mathrm{opp}}$ out of a Q-system in $\mathscr{C}$, using the $\alpha$-induction (Sect. 4.6) to extend $\rho \in \mathrm{End}(N)$ to $\alpha^{\pm}_{\rho} \in \mathrm{End}(M)$. In the simplest case, when the Q-system in $\mathscr{C}$ is trivial or Morita equivalent to the trivial Q-system, then one obtains the canonical Q-system Proposition 3.19 in $\mathscr{C} \boxtimes \mathscr{C}^{\mathrm{opp}}$. A more general analysis is given in [11, 12, 19].

**Proposition 4.37** *Let* **A** *be a Q-system in a braided C* tensor category. The full centre* $Z[\mathbf{A}]$ *is irreducible iff* **A** *is simple, i.e., iff the extension described by* **A** *is a factor* (*cf. Corollary* 3.40). *More generally, the following linear spaces have equal dimension:*

(i) $\mathrm{Hom}(\mathrm{id} \otimes \mathrm{id}, Z[\mathbf{A}])$
(ii) $\mathrm{Hom}(\mathrm{id}, C^{+}[\mathbf{A}])$ *and* $\mathrm{Hom}(\mathrm{id}, C^{-}[\mathbf{A}])$
(iii) *The centre* $M' \cap M$ *of the extension described by* **A**.

*Proof* The projection defining the full centre is a multiple of

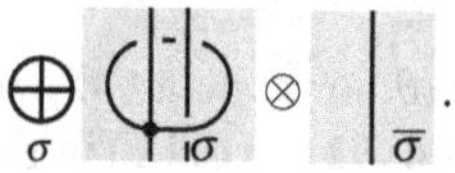

Therefore, for the multiplicity of the identity in $Z[\mathbf{A}]$, we can replace the canonical Q-system **R** by the trivial Q-system $\mathrm{id} \otimes \mathrm{id}$ in $\mathscr{C} \boxtimes \mathscr{C}^{\mathrm{opp}}$. Then trivially, $\dim\mathrm{Hom}(\mathrm{id} \otimes \mathrm{id}, Z[\mathbf{A}]) = \dim\mathrm{Hom}(\mathrm{id} \otimes \mathrm{id}, C^{+}[\mathbf{A} \otimes \mathrm{id}]) = \dim\mathrm{Hom}(\mathrm{id}, C^{+}[\mathbf{A}])$. Writing the centre projection as $p^{+}_{\mathbf{A}} = SS^*$, we have $t \in \mathrm{Hom}(\mathrm{id}, C^{+}[\mathbf{A}]) \subset \mathrm{Hom}(\mathrm{id}, \theta_P)$ iff $tn = \theta_P(n)t = S^*\theta(n)St$ iff $q = St \in \mathrm{Hom}(\mathrm{id}, \theta)$ satisfies $qn = P\theta(n)q = Pqn$ for all $n \in N$, i.e., $q = Pq$. Then Lemma 3.16(iii) together with the following Lemma prove the claim. □

**Lemma 4.38** *Let* **A** *be a Q-system in a braided C* tensor category, and* $P^{\pm} \equiv p^{\pm}_{\mathbf{A}}$ *its centre projections. Then* $q \in \mathrm{Hom}(\mathrm{id}, \theta)$ *satisfies* $qP^{+} = q$ *iff* $qP^{-} = q$ *iff* $q$ *satisfies Eq.* (3.2.4).

*Proof* We have

$qP^{+} = d_{\mathbf{A}}^{-1} \cdot$  $= d_{\mathbf{A}}^{-1} \cdot$  $= d_{\mathbf{A}}^{-1} \cdot$  $= qP^{-}.$

If $qP^{\pm} = q$, then $q$ satisfies Eq. (3.2.4) (using associativity):

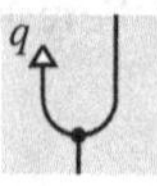 $= d_{\mathbf{A}}^{-1} \cdot$  $= d_{\mathbf{A}}^{-1} \cdot$ 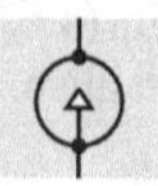 $= d_{\mathbf{A}}^{-1} \cdot$  $=$ 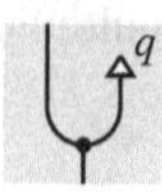 .

Conversely, if $q$ satisfies Eq. (3.2.4), then

$qP^+ = d_{\mathbf{A}}^{-1} \cdot$  $= d_{\mathbf{A}}^{-1} \cdot$  $=$  $= q.$

□

## 4.11 Modular Tensor Categories

A C* tensor category with finitely many inequivalent irreducible objects (denoted $\rho, \sigma, \tau$, etc.), all of finite dimension, is called *rational*. In a *braided* rational C* tensor category, one can introduce the matrices

$S_{\sigma,\tau} := \dim(\mathscr{C})^{-\frac{1}{2}} \cdot$ 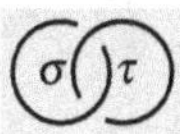 $= S_{\tau,\sigma}, \quad T^0_{\sigma,\tau} := \dfrac{\delta_{\sigma,\tau}}{\dim(\tau)} \cdot$  $\equiv \delta_{\sigma,\tau} \cdot \kappa_\tau,$

where $\dim(\mathscr{C}) = \sum_\rho \dim(\rho)^2$ is the global dimension Eq. (3.0.1), and $\kappa_\tau$ is the twist (Definition 4.18).

**Definition 4.39** A braiding of a tensor category $\mathscr{C}$ is called **non-degenerate** if there is no nontrivial sector $[\rho]$ such that $\varepsilon^+_{\rho,\sigma} = \varepsilon^-_{\rho,\sigma}$ for all $\sigma \in \mathscr{C}$. A braided rational C* tensor category is called **modular**, if the symmetric matrix $S$ is invertible.

**Proposition 4.40** ([20]) *A braided rational C* tensor category is modular if and only if it is non-degenerate. In this case, the matrix $S$ is unitary, and there is a complex phase $\omega$ (unique up to a third root of unity) such that the matrices $S$ and $T := \omega \cdot T^0$ form a unitary representation of the modular group $SL(2,\mathbb{Z})$:*

$$(ST^{-1})^3 = S^2, \quad S^4 = E.$$

*Moreover, $S_{\sigma,\bar\tau} = \overline{S_{\sigma,\tau}} = S_{\bar\sigma,\tau}$, i.e., the central element $S^2$ of $SL(2,\mathbb{Z})$ is represented by the conjugation matrix $C$.*

Recall that $\dim(\mathscr{C})^{\frac{1}{2}} = d_{\mathbf{R}}$ is also the dimension of the canonical Q-system in $\mathscr{C} \boxtimes \mathscr{C}^{\mathrm{opp}}$ (Proposition 3.19). By considering the id-id-component of the equation $T^{-1}ST^{-1}ST^{-1} = S$, one finds that $\omega^3 = \sum_\tau \kappa_\tau^{-1} \dim(\tau)^2 / d_{\mathbf{R}}$.

All the braidings mentioned in Example 4.19 are non-degenerate, giving rise to eight inequivalent modular categories associated with the same "fusion rules" of three irreducible sectors.

**Lemma 4.41** *For $\tau$ and $\sigma$ irreducible, one has in a modular category*

$\mathrm{RTr}_\sigma(\varepsilon_{\sigma,\tau}\varepsilon_{\tau,\sigma}) \equiv$ 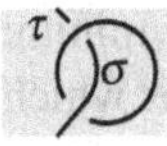 $= \dfrac{d_{\mathbf{R}} \cdot S_{\tau,\sigma}}{\dim(\tau)} \cdot 1_\tau = \mathrm{LTr}_\sigma(\varepsilon^*_{\bar\sigma,\tau}\varepsilon^*_{\tau,\bar\sigma}).$

*Proof* Clearly, $\mathrm{RTr}_\sigma(\varepsilon_{\sigma,\tau}\varepsilon_{\tau,\sigma})$ is a multiple of $1_\tau$. Thus, one can compute the coefficient by applying $\mathrm{Tr}_\tau$, where $\mathrm{Tr}_\tau(1_\tau) = \dim(\tau)$. Similar for the second equation. □

**Proposition 4.42** (The "killing ring") *For $\rho$ an object of $\mathcal{C}$, consider $\rho \otimes \mathrm{id}$ as an object of $\mathcal{C} \boxtimes \mathcal{C}^{\mathrm{opp}}$. If $\mathcal{C}$ is modular, then*

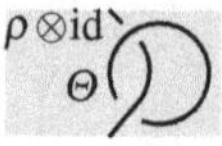

$$= d_{\mathbf{R}}^2 \cdot E_{\mathrm{id}} =$$

,

*where $\Theta$ is the endomorphism of the canonical Q-system, Corollary 3.21, and* $E_{\mathrm{id}} =$  $\in \mathrm{Hom}(\rho,\rho)$ *is the projection on the identity component* $\mathrm{id} \prec \rho$ *(which is zero if* id *is not contained in* $\rho$*).*

*Proof* If $\tau$ is irreducible, then

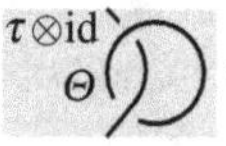

$$= \sum_\sigma$$
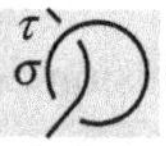

$$\otimes$$
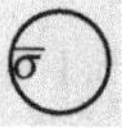

$$= \sum_\sigma \frac{d_{\mathbf{R}} \cdot S_{\tau,\sigma}}{\dim(\tau)} \cdot 1_\tau \cdot \dim(\sigma).$$

Then, $\dim(\sigma) = d_{\mathbf{R}} \cdot S_{\sigma,\mathrm{id}} = d_{\mathbf{R}} \cdot \overline{S_{\sigma,\mathrm{id}}}$ and unitarity of $S$ yield $d_{\mathbf{R}}^2$ if $\tau = \mathrm{id}$ and zero otherwise. If $\rho$ is reducible, then write $1_\rho = \sum_\tau E_\tau$ where $E_\tau \in \mathrm{Hom}(\rho,\rho)$ are the projections on the irreducible $\tau \prec \rho$. Under the "killing ring", only $\tau = \mathrm{id}$ survives. □

We can now see that it was essential to choose matching signs in the definition Definition 4.34 of the full centre:

**Corollary 4.43** *For* **A** *an irreducible Q-system in $\mathcal{C}$ and* **R** *the canonical Q-system, one has*

$$C^-[(\mathbf{A} \otimes \mathbf{1}) \times^+ \mathbf{R}] = C^+[(\mathbf{A} \otimes \mathbf{1}) \times^- \mathbf{R}] = \mathbf{R}.$$

*Proof* By using Proposition 4.42 and the fact that **R** is commutative, one can compute the trace $\mathrm{Tr}(p^\pm)$ of the respective centre projections of $(\mathbf{A} \otimes \mathbf{1}) \times^\mp \mathbf{R}$. The result is $\mathrm{Tr}(p^\pm) = d_{\mathbf{R}}^2$. On the other hand, the projection $p_{\mathbf{R}}$ onto the intermediate Q-system $\mathbf{R} = (\mathbf{1} \otimes \mathbf{1}) \times^\mp \mathbf{R} \prec (\mathbf{A} \otimes \mathbf{1}) \times^\mp \mathbf{R}$ satisfies Eq. (4.8.1), hence $p_{\mathbf{R}} < p^\pm$ by Proposition 4.28. Since by Proposition 2.4, $\mathrm{Tr}(p_{\mathbf{R}}) = \dim(\Theta_{\mathrm{can}}) = d_{\mathbf{R}}^2$, the claim follows. □

## 4.12 The Braided Product of Two Full Centres

We assume $\mathcal{C}$ to be modular.

The following Theorem 4.44 provides the minimal central projections for the braided product of two commutative Q-systems which arise as full centres. By way of preparation of this result, let us compile several equivalent ways of describing the centre.

Recall that the centre $M' \cap M$ of the extension corresponding to the braided product of two commutative Q-systems equals the relative commutant $\iota(N)' \cap M = \iota(\mathrm{Hom}(\Theta^{\mathbf{A}}\Theta^{\mathbf{B}}, \mathrm{id}))V$ by Lemma 3.16 and Proposition 4.33. The space $\mathrm{Hom}(\Theta^{\mathbf{A}}\Theta^{\mathbf{B}}, \mathrm{id})$ is isomorphic to $\mathrm{Hom}(\Theta^{\mathbf{B}}, \Theta^{\mathbf{A}})$ by Frobenius reciprocity. Thus, there is a linear bijection

$$\chi : \mathrm{Hom}(\Theta^{\mathbf{B}},\Theta^{\mathbf{A}}) \to M'\cap M, \quad \chi(T) := \iota\left(R^{\mathbf{A}*}\circ(1_{\Theta^{\mathbf{A}}}\times T)\right)V =$$

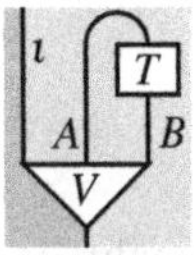


(4.12.1)

with inverse

$$\chi^{-1}(\cdot) = [1_{\Theta^{\mathbf{A}}} \times (W^* \circ \bar{\iota}(\cdot))] \circ R^{\mathbf{A}}.$$

Notice also that $\bar{\iota}$ maps the centre into $\mathrm{Hom}(\Theta^{\mathbf{A}}\Theta^{\mathbf{B}}, \Theta^{\mathbf{A}}\Theta^{\mathbf{B}})$:

$$\bar{\iota}\chi(T) = \left(1_{\Theta^{\mathbf{A}}\Theta^{\mathbf{B}}} \times \left(R^{\mathbf{A}*}\circ(1_{\Theta^{\mathbf{A}}}\times T)\right)\right)\circ X =$$

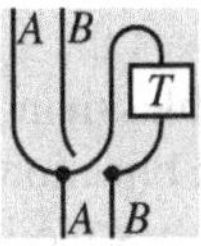


$\overset{(4.5.1)}{=}$

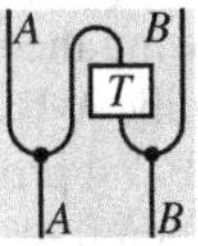


$\equiv$

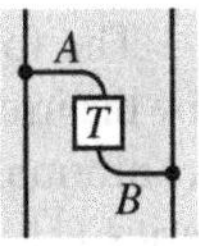


where we have used commutativity of **B**, and are freely appealing to Frobenius reciprocity in the last way of drawing the diagram.

Then, one easily verifies that

$$\chi(T_1)\circ\chi(T_2) = \chi(T_1 * T_2), \qquad \bar{\iota}\chi(T_1)\circ\bar{\iota}\chi(T_2) = \bar{\iota}\chi(T_1 * T_2),$$

where $T_1 * T_1$ is the commutative "convolution" product on $\mathrm{Hom}(\Theta^{\mathbf{B}}, \Theta^{\mathbf{A}})$ with unit $W^{\mathbf{A}}W^{\mathbf{B}*}$:

$T_1 * T_2 :=$

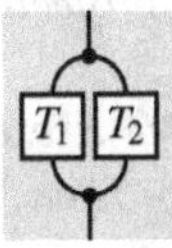


$\overset{(4.5.1)}{=}$

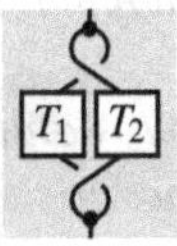


$=$

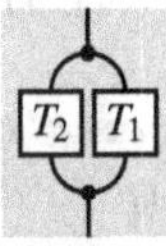


,

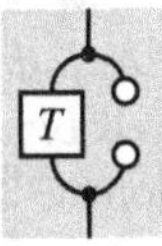


$=$

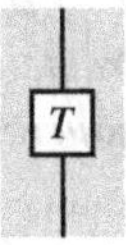


$=$

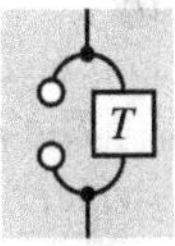


.

(4.12.2)

Likewise, the adjoint is given by

$$\chi(T)^* = \chi(F(T)), \qquad \bar{\iota}\chi(T)^* = \bar{\iota}\chi(F(T)),$$

where $F$ is the antilinear Frobenius conjugation on $\mathrm{Hom}(\Theta^{\mathbf{B}}, \Theta^{\mathbf{A}})$

$F(T) =$

$\overset{\text{Prop. 2.6}}{=}$

$\in \mathrm{Hom}(\Theta^{\mathbf{B}},\Theta^{\mathbf{A}}).$

Therefore, finding the minimal projections $E_m \in M' \cap M$ is equivalent to finding the minimal projections $I_m \in \mathrm{Hom}(\Theta^{\mathbf{B}}, \Theta^{\mathbf{A}})$ w.r.t. the convolution product, i.e., to solving the system

$$\begin{aligned} \text{self-adjointness } & I_m^* = F(I_m), \\ \text{idempotency } & I_m * I_m' = \delta_{mm'} \cdot I_m, \\ \text{completeness } & \sum_m I_m = W^{\mathbf{A}} W^{\mathbf{B}*}. \end{aligned} \tag{4.12.3}$$

Minimality is ensured if the number of $I_m$ exhausts the dimension of $\mathrm{Hom}(\Theta^{\mathbf{B}}, \Theta^{\mathbf{A}})$. We therefore have to solve these equations by a basis $I_m$ of $\mathrm{Hom}(\Theta^{\mathbf{B}}, \Theta^{\mathbf{A}})$, and put $E_m = \chi(I_m)$. Obviously, then also $P_m = \bar{\iota}(E_m) =$ 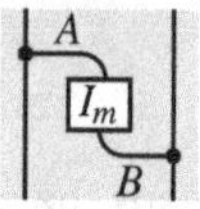

$\in \mathrm{Hom}(\Theta^{\mathbf{A}}\Theta^{\mathbf{B}}, \Theta^{\mathbf{A}}\Theta^{\mathbf{B}})$ will be projections.

The following theorem gives the solution to Eq. (4.12.3), where $I_{\mathbf{m}}$ are labelled by the irreducibe **A**-**B**-bimodules **m** in $\mathcal{C}$. This result is of great interest for boundary conformal QFT: it provides a bijection between chiral bimodules and phase boundaries [1]. It therefore establishes the link between our algebraic QFT approach to phase boundaries, and the TFT approach by [21–24]. The fact that the central projections for the braided product extension of two full centre Q-systems in $\mathcal{C} \boxtimes \mathcal{C}^{\mathrm{opp}}$ are labelled by bimodules in $\mathcal{C}$, means in physical terms that the boundary conditions between two maximal local two-dimensional extensions is fixed by chiral data.

**Theorem 4.44** *Let* **A** *and* **B** *be two simple Q-systems in a modular tensor category $\mathcal{C}$, and let $\mathrm{Z}[\mathbf{A}] = (\Theta^{\mathbf{A}}, W^{\mathbf{A}}, X^{\mathbf{A}})$ and $\mathrm{Z}[\mathbf{B}] = (\Theta^{\mathbf{B}}, W^{\mathbf{B}}, X^{\mathbf{B}})$ be their full centre Q-systems in $\mathcal{C} \boxtimes \mathcal{C}^{\mathrm{opp}}$. Let $N \subset M$ be the extension defined by either of the product Q-systems $\mathrm{Z}[\mathbf{A}] \times^{\pm} \mathrm{Z}[\mathbf{B}]$. Then $M$ has a centre given by $M' \cap M = \iota(N)' \cap M = \mathrm{Hom}(\iota, \iota)$. The minimal central projections $E_{\mathbf{m}}$ can be characterized as follows.*

*The irreducible* **A**-**B**-*bimodules* $\mathbf{m} = (\beta, m)$ *naturally give rise to intertwiners $D_{R[\mathbf{m}]\|Z} \in \mathrm{Hom}(\Theta^{\mathbf{B}}, \Theta^{\mathbf{A}})$ (to be defined in the proof). Then*

$$I_{\mathbf{m}} = \frac{\dim(\beta)}{d_{\mathbf{A}}^2 d_{\mathbf{B}}^2 d_{\mathbf{R}}^2} \cdot D_{R[\mathbf{m}]\|Z}$$

*solve the system Eq. (4.12.3). Then $E_{\mathbf{m}} = \chi(I_{\mathbf{m}})$ are the minimal central projections.*

As a byproduct, we shall also prove:

**Proposition 4.45** *Let* **A** *be a simple Q-system in a modular tensor category, so that its centre* $\mathrm{Z}[\mathbf{A}]$ *is irreducible (Proposition 4.37). Then* $d_{\mathrm{Z}[\mathbf{A}]} = d_{\mathbf{R}} = \dim(\mathcal{C})^{\frac{1}{2}}$

*equals the dimension of the canonical Q-system. In particular, all irreducible full centres have the same dimension.*

(This is not a new result, cf. [25], but the proof seems to be new.)

The proof of Theorem 4.44 is rather lengthy, but it is worthwhile because it paves the way to an efficient computation of the centre, with ensuing classification results. The operators $I_{\mathbf{m}}$ first appeared in [16], but their idempotent property is not manifest there. It was proven in a more special case in [25] (with the hindsight that the general case can be reduced to the special case by highly nontrivial properties of modular tensor categories). We attempt to give here a streamlined version of the proof that does not require the general theory of modular tensor categories. The use of the C*-structure of the DHR category allows for some substantial simplification as compared to [25].

*Proof of Theorem* 4.44 The statement about the centre is just an instance of Lemma 3.16 and Proposition 4.33, because the full centres are commutative.

To prepare the solution of Eq. (4.12.3), we associate with every **A**-**B**-bimodule $\mathbf{m} = (\beta, m \in \mathrm{Hom}(\beta, \theta^{\mathbf{A}}\beta\theta^{\mathbf{B}}))$ an intertwiner $D_{\mathbf{m}} \in \mathrm{Hom}(\theta^{\mathbf{B}}, \theta^{\mathbf{A}})$ as follows [25]:

$$D_{\mathbf{m}} := \mathrm{Tr}_\beta\left(\varepsilon_{\theta^{\mathbf{A}},\beta} \circ (1_{\theta^{\mathbf{A}}\beta} \times r^{\mathbf{B}*}) \circ (m \times 1_{\theta^{\mathbf{B}}})\right):$$

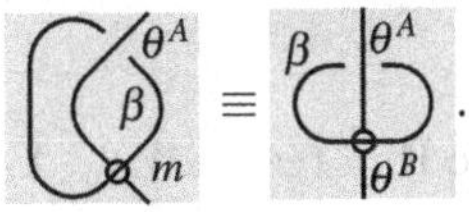


(We freely use Frobenius reciprocity in the graphical representations.) One easily sees

**Lemma 4.46** (cf. [25]) *The following statements hold.*

(i) $D_{\mathbf{m}}$ *depends only on the unitary equivalence class of* $\mathbf{m} = (\beta, m)$.
(ii) $D^*_{\mathbf{m}} = D_{\overline{\mathbf{m}}}$.
(iii) $D_{\mathbf{m}_1\oplus\mathbf{m}_2} = D_{\mathbf{m}_1} + D_{\mathbf{m}_2}$.
(iv) *If* $\mathbf{m} = (\beta, m)$ *is an* **A**-**B***-bimodule and* $\mathbf{m}' = (\beta', m')$ *a* **B**-**C***-bimodule, hence* $\mathbf{m} \otimes_{\mathbf{B}} \mathbf{m}'$ *an* **A**-**C***-bimodule, then* $D_{\mathbf{m}} D_{\mathbf{m}'} = d_{\mathbf{B}} \cdot D_{\mathbf{m}\otimes_{\mathbf{B}}\mathbf{m}'}$.
(v) $w^{\mathbf{A}*} D_{\mathbf{m}} w^{\mathbf{B}} = \dim(\mathbf{m}) \equiv \dim(\beta)$ *for* $\mathbf{m} = (\beta, m)$.

*Proof* (i) follows because the "closed $\beta$-line" represents a trace, absorbing a unitary bimodule morphism : $\mathbf{m} \to \mathbf{m}'$. (ii) is proven in the same way as Lemma 4.26, using the unitarity of the twist. (iii) follows by

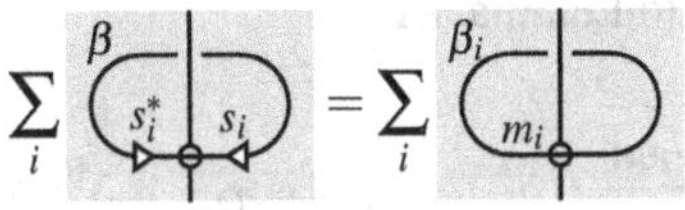


(iv) follows from

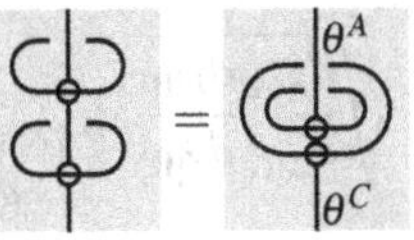

in combination with the property Eq. (3.7.1) in Lemma 3.36. (v) follows from the unit property and Eq. (2.2.6). □

In particular, taking $\mathbf{A} \equiv (\theta^{\mathbf{A}}, x^{\mathbf{A}(2)})$ as the trivial **A**-**A**-bimodule, one has

**Corollary 4.47** (i) $D_{\mathbf{A}} = d_{\mathbf{A}} \cdot P^l_{\mathbf{A}}$, *hence (by Lemma* 4.46(iv))
(ii) $P^l_{\mathbf{A}} \cdot D_{\mathbf{m}} \cdot P^l_{\mathbf{B}} = D_{\mathbf{m}}$.

We also define more generally, for any $\rho \in \mathscr{C}$,

$$D_{\mathbf{m}}(\rho) :=$$ 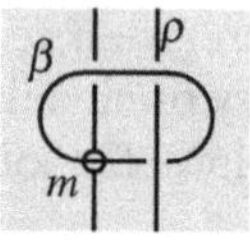 $$\in \mathrm{Hom}(\theta^{\mathbf{B}} \circ \rho, \theta^{\mathbf{A}} \circ \rho).$$


Clearly, $D_{\mathbf{m}} \equiv D_{\mathbf{m}}(\mathrm{id})$. The properties (i)–(iv) hold as well for $D_{\mathbf{m}}(\rho)$.

Next, consider $\mathbf{A} \otimes \mathbf{1} \equiv (\theta \otimes \mathrm{id}, x \otimes 1_{\mathrm{id}})$ as a Q-system in $\mathscr{C} \boxtimes \mathscr{C}^{\mathrm{opp}}$, and $\mathbf{m} \otimes \mathbf{1} \equiv (\beta \otimes \mathrm{id}, m \otimes 1_{\mathrm{id}})$ as an $\mathbf{A} \otimes \mathbf{1}$-$\mathbf{B} \otimes \mathbf{1}$-bimodule. Taking the product

$$R[\mathbf{A}] := (\mathbf{A} \otimes \mathbf{1}) \times^{+} \mathbf{R}$$

where $\mathbf{R} = (\Theta_{\mathrm{can}}, W_{\mathrm{can}}, X_{\mathrm{can}})$ is the canonical Q-system in $\mathscr{C} \boxtimes \mathscr{C}^{\mathrm{opp}}$, we get

$$R[\mathbf{m}] = ((\beta \otimes \mathrm{id}) \circ \Theta_{\mathrm{can}}, R[\mathbf{m}]), \qquad R[\mathbf{m}] =$$ 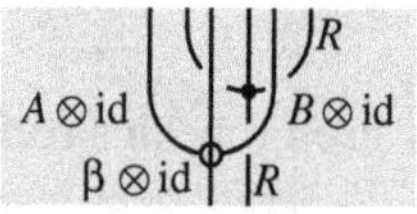


as an $R[\mathbf{A}]$-$R[\mathbf{B}]$-bimodule. Because **R** is commutative, one has $D_{\mathbf{R}} = d_{\mathbf{R}} \cdot 1_{\Theta_{\mathrm{can}}}$ by Corollary 4.47, and hence

$$D_{R[\mathbf{m}]} =$$ 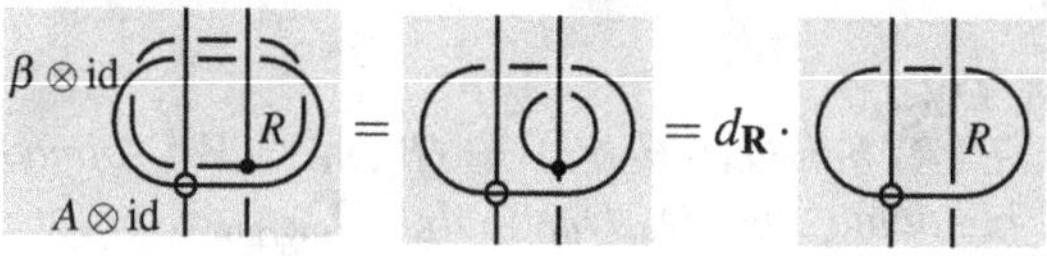 $$= \ldots = d_{\mathbf{R}} \cdot \ldots = d_{\mathbf{R}} \cdot D_{\mathbf{m}\otimes\mathbf{1}}(\Theta_{\mathrm{can}}).$$


(4.12.4)

As the full centre $Z[\mathbf{A}] = (\Theta^{\mathbf{A}}, W^{\mathbf{A}}, X^{\mathbf{A}})$ is an irreducible (by Proposition 4.37) intermediate Q-system to $R[\mathbf{A}]$, the $R[\mathbf{A}]$-$R[\mathbf{B}]$-bimodule $R[\mathbf{m}]$ restricts to a $Z[\mathbf{A}]$-$Z[\mathbf{B}]$-bimodule according to Lemma 4.15

$$R[\mathbf{m}]|_Z = \sqrt{\frac{d_{R[\mathbf{A}]} d_{R[\mathbf{B}]}}{d_{Z[\mathbf{A}]} d_{Z[\mathbf{B}]}}} \cdot (S^{\mathbf{A}*} \times 1_{(\beta \otimes \mathrm{id}) \times \Theta_{\mathrm{can}}} \times S^{\mathbf{B}}) \circ R[\mathbf{m}] \equiv$$

$$\equiv \sqrt{\frac{d_{R[\mathbf{A}]} d_{R[\mathbf{B}]}}{d_{Z[\mathbf{A}]} d_{Z[\mathbf{B}]}}} \cdot$$ 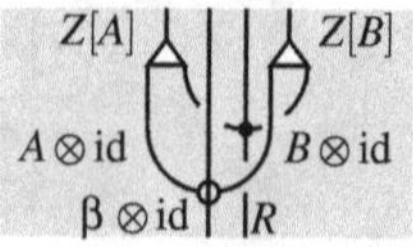 ,


(4.12.5)

where $S^{\mathbf{A}} \in \mathrm{Hom}(Z[\mathbf{A}], R[\mathbf{A}])$, $S^{\mathbf{B}} \in \mathrm{Hom}(Z[\mathbf{B}], R[\mathbf{B}])$ are isometric intertwiners such that $S^{\mathbf{A}}S^{\mathbf{A}*} = P^l_{R[\mathbf{A}]}$, $S^{\mathbf{B}}S^{\mathbf{B}*} = P^l_{R[\mathbf{B}]}$, cf. Lemma 4.15.

Next, we consider the intertwiners

$$D_{R[\mathbf{m}]\|Z} \in \mathrm{Hom}(\Theta^{\mathbf{B}}, \Theta^{\mathbf{A}}).$$

In particular, for the trivial **A**-**A**-bimodule **A**, one has, using Corollary 4.47

$$D_{R[\mathbf{A}]\|Z} = \frac{d_{R[\mathbf{A}]}}{d_{Z[\mathbf{A}]}} \cdot S^* D_{R[\mathbf{A}]} S = \frac{d^2_{R[\mathbf{A}]}}{d_{Z[\mathbf{A}]}} \cdot S^* P^l_{R[\mathbf{A}]} S = \frac{d^2_{\mathbf{A}} d^2_{\mathbf{R}}}{d_{Z[\mathbf{A}]}} \cdot 1_{\Theta^{\mathbf{A}}}. \qquad (4.12.6)$$

We introduce the positive-definite inner product on $\mathrm{Hom}(\Theta^{\mathbf{B}}, \Theta^{\mathbf{A}})$ w.r.t. the trace:

$$(T_2, T_1) := \mathrm{Tr}_{\Theta^{\mathbf{B}}}(T_1^* T_2) =$$ 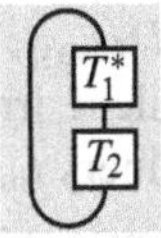

$$\equiv \mathrm{Tr}_{\Theta^{\mathbf{A}}}(T_2 T_1^*) =$$ 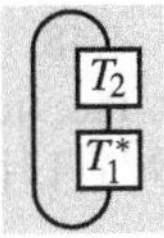

. (4.12.7)

Then we compute

$$(D_{R[\mathbf{m}']\|Z}, D_{R[\mathbf{m}]\|Z}) = \frac{d_{R[\mathbf{A}]} d_{R[\mathbf{B}]}}{d_{Z[\mathbf{A}]} d_{Z[\mathbf{B}]}} \cdot \mathrm{Tr}_{\Theta^{\mathbf{A}}}(S^* D_{R[\mathbf{m}']} S S^* D_{R[\overline{\mathbf{m}}]} S) \qquad (4.12.8)$$

$$\overset{\text{Corollary 4.47, (4.12.4)}}{=} \frac{d_{R[\mathbf{A}]} d_{R[\mathbf{B}]}}{d_{Z[\mathbf{A}]} d_{Z[\mathbf{B}]}} \cdot d^2_{\mathbf{R}} \cdot \mathrm{Tr}_{(\theta^{\mathbf{A}} \otimes \mathrm{id})\Theta_{\mathrm{can}}}(D_{\mathbf{m}' \otimes \mathbf{1}}(\Theta_{\mathrm{can}}) D_{\overline{\mathbf{m}} \otimes \mathbf{1}}(\Theta_{\mathrm{can}}))$$

$$\overset{\text{Lemma 4.46 (iii),(iv)}}{=} \frac{d_{R[\mathbf{A}]} d_{R[\mathbf{B}]}}{d_{Z[\mathbf{A}]} d_{Z[\mathbf{B}]}} \cdot d^2_{\mathbf{R}} \cdot d_{\mathbf{B}} \sum_{\mathbf{n}} N^{\mathbf{n}}_{\mathbf{m}'\overline{\mathbf{m}}} \cdot \mathrm{LTr}_{\theta^{\mathbf{A}} \otimes \mathrm{id}} \, \mathrm{RTr}_{\Theta_{\mathrm{can}}}(D_{\mathbf{n} \otimes \mathbf{1}}(\Theta_{\mathrm{can}})),$$

where $\mathbf{m}' \otimes_{\mathbf{B}} \overline{\mathbf{m}} \simeq \bigoplus_{\mathbf{n}} N^{\mathbf{n}}_{\mathbf{m}'\overline{\mathbf{m}}} \cdot \mathbf{n}$ as **A**-**A**-bimodules, $\mathbf{n} = (\alpha, n)$ irreducible.

At this point, modularity comes to bear through Proposition 4.42: namely $\mathrm{RTr}_{\Theta_{\mathrm{can}}}$ projects on the contribution $\mathrm{id} \prec \alpha$:

$$\mathrm{RTr}_{\Theta_{\mathrm{can}}}(D_{\mathbf{n} \otimes \mathbf{1}}(\Theta_{\mathrm{can}})) =$$ 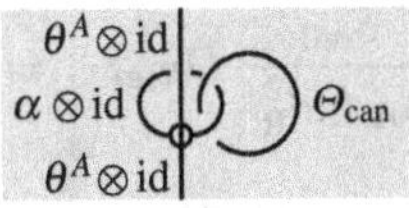

$$= d^2_{\mathbf{R}} \cdot$$ 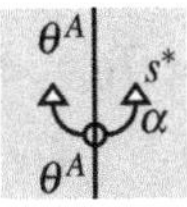

$\otimes \mathrm{id}$,

where $s$ is an isometry such that $ss^* = E_{\mathrm{id}} \in \mathrm{Hom}(\alpha, \alpha)$. After taking also $\mathrm{LTr}_{\theta^{\mathbf{A}} \otimes \mathrm{id}}$, one obtains

$$\mathrm{LTr}_{\theta^{\mathbf{A}} \otimes \mathrm{id}} \, \mathrm{RTr}_{\Theta_{\mathrm{can}}}(D_{\mathbf{n} \otimes \mathbf{1}}(\Theta_{\mathrm{can}})) = d^2_{\mathbf{R}} \cdot$$ 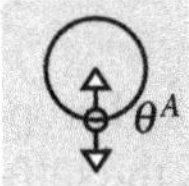

Now, Lemma 3.41 applies, and accordingly the value vanishes unless $\mathbf{n}$ is the trivial $\mathbf{A}$-$\mathbf{A}$-bimodule $\mathbf{n} = \mathbf{A}$. In this case $s = w/\sqrt{d_\mathbf{A}}$, hence the value of the diagram is $\dim(\theta)/d_\mathbf{A} = d_\mathbf{A}$. Since $N^\mathbf{A}_{\mathbf{m}'\overline{\mathbf{m}}} = \delta_{\mathbf{mm}'}$, we arrive at the orthogonality relation

**Corollary 4.48** *In a modular category,*

$$(D_{R[\mathbf{m}']\|Z}, D_{R[\mathbf{m}]\|Z}) = \frac{d_\mathbf{A}^2 d_\mathbf{B}^2 d_\mathbf{R}^6}{d_{Z[\mathbf{A}]} d_{Z[\mathbf{B}]}} \cdot \delta_{\mathbf{mm}'}.$$

For $\mathbf{m} = \mathbf{A}$, $D_{R[\mathbf{A}]\|Z} = D^*_{R[\mathbf{A}]\|Z}$, one can compute $(D_{R[\mathbf{A}]\|Z}, D_{R[\mathbf{A}]\|Z})$ in two different ways: By Eq. (4.12.6), the result is $d_\mathbf{A}^4 d_\mathbf{R}^4$. By Corollary 4.48, it is $\frac{d_\mathbf{A}^2 d_\mathbf{B}^2 d_\mathbf{R}^6}{d_{Z[\mathbf{A}]} d_{Z[\mathbf{B}]}}$. By comparison, $d_{Z[\mathbf{A}]} = d_\mathbf{A}$ for all simple Q-systems $\mathbf{A}$. This proves Proposition 4.45. Moreover, the coefficient in Corollary 4.48 equals $d_\mathbf{A}^2 d_\mathbf{B}^2 d_\mathbf{R}^4$.

By Corollary 4.48, the intertwiners $D_{R[\mathbf{m}]\|Z}$ are linearly independent. It is also known that the number of inequivalent irreducible $\mathbf{A}$-$\mathbf{B}$-bimodules equals $\dim \mathrm{Hom}(\Theta^\mathbf{A}, \Theta^\mathbf{B})$ [13], hence the intertwiners $I_\mathbf{m}$ span $\mathrm{Hom}(\Theta^\mathbf{A}, \Theta^\mathbf{B})$. Since both $(d_\mathbf{A} d_\mathbf{B} d_\mathbf{R}^2)^{-1} \cdot D_{R[\mathbf{m}]\|Z}$ and $T := (\dim(\sigma)\dim(\tau))^{-\frac{1}{2}} \cdot T^\mathbf{A}_{\sigma\otimes\tau} T^{\mathbf{B}*}_{\sigma\otimes\tau}$ (where $T^\mathbf{A}$ and $T^\mathbf{B}$ are isometric bases of $\mathrm{Hom}(\sigma \otimes \tau, \Theta^\mathbf{A})$ resp. $\mathrm{Hom}(\sigma \otimes \tau, \Theta^\mathbf{B})$ for all irreducible common sub-endomorphism $\sigma \otimes \tau$ of $\Theta^\mathbf{A}$ and $\Theta^\mathbf{B}$) form orthonormal bases w.r.t. trace, the matrix

$$S_{\mathbf{m}T} := \frac{1}{d_\mathbf{A} d_\mathbf{B} d_\mathbf{R}^2 \sqrt{\dim(\sigma \otimes \tau)}} \cdot \mathrm{Tr}_{\Theta^\mathbf{A}}(D_{R[\mathbf{m}]\|Z} T^\mathbf{B}_{\sigma\otimes\tau} T^{\mathbf{A}*}_{\sigma\otimes\tau})$$

is unitary. In particular, for $\sigma = \tau = \mathrm{id}$, $T = T_0 \equiv d_\mathbf{R}^{-1} \cdot W^\mathbf{A} W^{\mathbf{B}*}$, one finds

$$S_{\mathbf{m}0} = \frac{\dim(\beta)}{d_\mathbf{A} d_\mathbf{B}},$$

hence

$$W^\mathbf{A} W^{\mathbf{B}*} = d_\mathbf{R} \cdot T_0 = d_\mathbf{R} \sum_\mathbf{m} \frac{S_{\mathbf{m}0}}{d_\mathbf{A} d_\mathbf{B} d_\mathbf{R}^2} \cdot D_{R[\mathbf{m}]\|Z} = \sum_\mathbf{m} \frac{\dim(\beta)}{d_\mathbf{A}^2 d_\mathbf{B}^2 d_\mathbf{R}^2} \cdot D_{R[\mathbf{m}]\|Z}. \tag{4.12.9}$$

Now, we define

$$I_\mathbf{m} := \frac{\dim(\beta)}{d_\mathbf{A}^2 d_\mathbf{B}^2 d_\mathbf{R}^2} \cdot D_{R[\mathbf{m}]\|Z} \in \mathrm{Hom}(\Theta^\mathbf{B}, \Theta^\mathbf{A}). \tag{4.12.10}$$

From the definition and properties (i) and (ii) in Lemma 4.46, one can see that $D_{R[\mathbf{m}]\|Z}$ and hence $I_\mathbf{m}$ satisfy the selfadjointness condition in Eq. (4.12.3). Because $W^\mathbf{A} W^{\mathbf{B}*}$ is the unit w.r.t. the convolution product Eqs. (4.12.2), and (4.12.9) is the completeness relation, i.e., $\sum_\mathbf{m} E_\mathbf{m} = \mathbf{1}_M$ under the isomorphism $\chi$. It remains to prove the idempotency relation in Eq. (4.12.3). Using Eq. (4.12.9), it suffices to show that

$$I_{\mathbf{m}} * I_{\mathbf{m}'} = 0$$

for $\mathbf{m} \neq \mathbf{m}'$, in order to conclude that $I_{\mathbf{m}} * I_{\mathbf{m}} = I_{\mathbf{m}} * \sum_{\mathbf{m}'} I_{\mathbf{m}'} = I_{\mathbf{m}} * (W^{\mathbf{A}} W^{\mathbf{B}*}) = I_{\mathbf{m}}$.

Let $\mathbf{m}$ and $\mathbf{m}'$ be two $\mathbf{A}$-$\mathbf{B}$-bimodules. Define

$Q_{\mathbf{m},\mathbf{m}'} :=$ 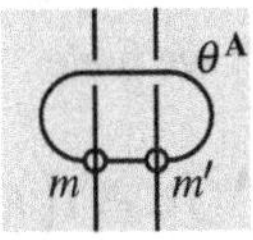

$\in \mathrm{Hom}(\beta\overline{\beta}', \beta\overline{\beta}')$.

By a similar computation as for the projection property of the left and right centre, Lemma 4.26, one sees that $(d_{\mathbf{A}} d_{\mathbf{B}})^{-1} \cdot Q_{\mathbf{m},\mathbf{m}'}$ is a projection. Now,

$$\mathrm{LTr}_{\beta}\,\mathrm{RTr}_{\overline{\beta}'}(Q_{\mathbf{m},\mathbf{m}'}) = \mathrm{Tr}_{\theta^{\mathbf{A}}}(D_{\mathbf{m}} D_{\mathbf{m}'}^{*}).$$

Thus, replacing $\mathbf{A}$ and $\mathbf{B}$ by $Z[\mathbf{A}]$ and $Z[\mathbf{B}]$, $\mathbf{m}$ and $\mathbf{m}'$ by $R[\mathbf{m}]|_Z$ and $R[\mathbf{m}']|_Z$, and $\beta$ and $\beta'$ by $R[\beta] = (\mathrm{id} \otimes \beta)\Theta_{\mathrm{can}}$ and $R[\beta']$, we conclude that

$$\mathrm{LTr}_{R[\beta]}\,\mathrm{RTr}_{R[\overline{\beta}']}(Q_{R[\mathbf{m}]|_Z, R[\mathbf{m}']|_Z}) = 0$$

for $\mathbf{m} \neq \mathbf{m}'$ by the orthogonality of $D_{R[\mathbf{m}]|_Z}$, Corollary 4.48. Since $Q_{R[\mathbf{m}]|_Z, R[\mathbf{m}']|_Z}$ is a multiple of a projection, hence a positive operator, and because the traces are faithful positive maps, it follows that $Q_{R[\mathbf{m}]|_Z, R[\mathbf{m}']|_Z} = 0$ for $\mathbf{m} \neq \mathbf{m}'$.

Now in order to conclude that $X^{\mathbf{A}*}(I_{\mathbf{m}} \times I_{\mathbf{m}'})X^{\mathbf{B}} = 0$ for $\mathbf{m} \neq \mathbf{m}'$, it suffices to compute

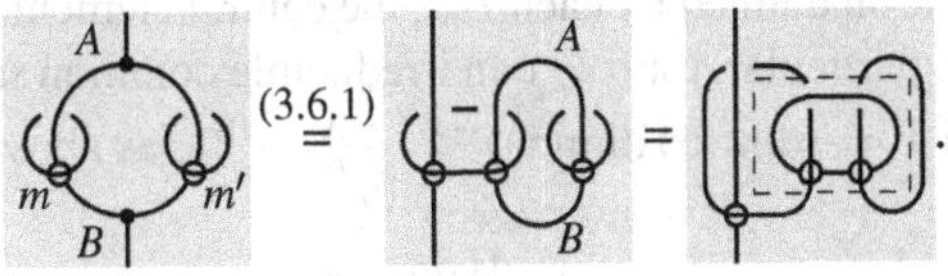


Inside the dashed box, there appears the intertwiner $Q_{R[\mathbf{m}]|_Z, R[\mathbf{m}']|_Z}$, which we have just shown to be zero if $\mathbf{m} \neq \mathbf{m}'$. In step (r), the representation property of $\mathbf{m}$ as a left $\mathbf{A}$-module has been used. This concludes the proof that $I_{\mathbf{m}}$ solve Eq. (4.12.3).

Theorem 4.44 now follows from the considerations before Eq. (4.12.3). □

The minimal projections $E_{\mathbf{m}} \in M' \cap M$ define representations $m \mapsto E_{\mathbf{m}} m$ as in Sect. 4.2. In these representations, the generators $V^{\mathbf{A}}$ and $V^{\mathbf{B}}$ of the intermediate algebras $M^{\mathbf{A}}$ and $M^{\mathbf{B}}$ (defined as in Lemma 4.31) are no longer independent. Let us describe the nature of these relations.

**Lemma 4.49** *The bijection* $\chi$, *Eq.* (4.12.1), *can be equivalently written as*

$$\chi(T) = V^{BA*}\iota(T)V^{\mathbf{B}}.$$

*Therefore, in particular,* $E_{\mathbf{m}} = V^{\mathbf{A}*}\iota(I_{\mathbf{m}})V^{\mathbf{B}}$.

*Proof* By Lemma 4.31 and $V^* = \iota(R^*)V$, we have

$$V^{BA*}\iota(T)V^{\mathbf{B}} = V^*\iota\Theta^{\mathbf{A}}(W^{\mathbf{B}})\iota(T)\iota(W^{\mathbf{A}*})V = \iota\Big(R^*\Theta(\Theta^{\mathbf{A}}(W^{\mathbf{B}})TW^{\mathbf{A}*})X\Big)V,$$

and the argument of $\iota$ equals

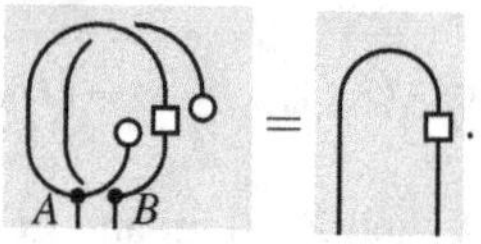


This coincides with Eq. (4.12.1). The last statement follows from $E_{\mathbf{m}} = \chi(I_{\mathbf{m}})$. □

Expanding a general element $T \in \mathrm{Hom}(\Theta^{\mathbf{B}}, \Theta^{\mathbf{A}})$ in the basis $I_{\mathbf{m}}$, such that $T = \sum_{\mathbf{m}} c_{\mathbf{m}}(T) \cdot I_{\mathbf{m}}$, we get

$$V^{\mathbf{A}*}\iota(T)V^{\mathbf{B}} = \sum_{\mathbf{m}} c_{\mathbf{m}}(T) \cdot E_{\mathbf{m}},$$

i.e., in the representation defined by each $E_{\mathbf{m}}$, the central elements $V^{\mathbf{A}*}\iota(T)V^{\mathbf{B}}$ take the values $c_{\mathbf{m}}(T)$. In particular, for $\sigma \otimes \tau$ an irreducible common sub-endomorphism of $\Theta^{\mathbf{A}}$ and $\Theta^{\mathbf{B}}$, and $T = (\dim(\sigma)\dim(\tau))^{-\frac{1}{2}} \cdot T^{\mathbf{A}}_{\sigma\otimes\tau}T^{\mathbf{B}*}_{\sigma\otimes\tau}$ as above, these values are

$$c_{\mathbf{m}}(T) = \frac{d_{\mathbf{A}}d_{\mathbf{B}}}{\dim(\beta)} \cdot \overline{S_{\mathbf{m}T}}.$$

Since on the other hand, the charged intertwiners $\Psi^{\mathbf{A}}_{\sigma\otimes\tau} = \iota(T^{\mathbf{A}*}_{\sigma\otimes\tau})V^{\mathbf{A}} \in \mathrm{Hom}(\iota^{\mathbf{A}}, \iota^{\mathbf{A}} \circ (\sigma \otimes \tau))$ and $\Psi^{\mathbf{B}}_{\sigma\otimes\tau} = \iota(T^{\mathbf{B}*}_{\sigma\otimes\tau})V^{\mathbf{B}}$ are multiples of isometries because $\iota^{\mathbf{A}}$ and $\iota^{\mathbf{B}}$ are irreducible, the numerical values for $\Psi^{\mathbf{A}*}_{\sigma\otimes\tau}\Psi^{\mathbf{B}}_{\sigma\otimes\tau}$ define "angles" between them [1].

*Example 4.50* Let **A** and **B** be the trivial Q-system (or Morita equivalent), such that the full centres coincide with **R**. The irreducible bimodules of the trivial Q-system are just the irreducible endomorphisms $\sigma \in \mathscr{C}$, $\mathbf{m} = (\sigma, 1_\sigma)$. The irreducible sub-endomorphism of $\Theta^{\mathbf{A}} = \Theta^{\mathbf{B}} = \Theta_{\mathrm{can}}$ are $\tau \otimes \overline{\tau}$. The operators $I_{\mathbf{m}}$, Eq. (4.12.10), simplify to $\frac{\dim(\sigma)}{d_{\mathbf{R}}} \cdot$

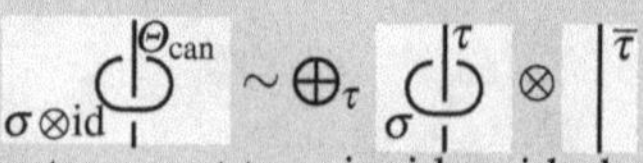


The matrix $(S_{\mathbf{m},T})_{\mathbf{m},T}$ determining the angles turns out to coincide with the "modular" matrix $(S_{\sigma,\tau})_{\sigma,\tau}$, cf. Definition 4.39. In particular, if $S_{\sigma,\tau}$ happens to equal a complex phase $\omega$ times $\dim(\sigma)\dim(\tau)\cdot\left(\sum_\rho \dim(\rho)^2\right)^{-\frac{1}{2}}$ (this is always the case whenever $\sigma$ has dimension $\dim(\sigma)=1$), it follows that the generators $\Psi^{\mathbf{A}}_{\tau\otimes\bar{\tau}} = \omega\cdot\Psi^{\mathbf{B}}_{\tau\otimes\bar{\tau}}$ are linearly dependent in the representation given by $E_{\mathbf{m}}$.

## References


1. M. Bischoff, Y. Kawahigashi, R. Longo, K.-H. Rehren, Phase boundaries in algebraic conformal QFT. arXiv:1405.7863
2. D. Bisch, A note on intermediate subfactors. Pac. J. Math. **163**, 201–216 (1994)
3. A. Bartels, C.L. Douglas, A. Henriques, Dualizability and index of subfactors. arXiv:1110.5671
4. R. Longo, J.E. Roberts, A theory of dimension. K-Theory **11**, 103–159 (1997) (notably Chaps. 3 and 4)
5. M. Müger, Galois theory for braided tensor categories and the modular closure. Adv. Math. **150**, 151–201 (2000)
6. R. Longo, K.-H. Rehren, Nets of subfactors. Rev. Math. Phys. **7**, 567–597 (1995)
7. J. Böckenhauer, D. Evans, Modular invariants, graphs and $\alpha$-induction for nets of subfactors, I. Commun. Math. Phys. **197**, 361–386 (1998)
8. J. Böckenhauer, D. Evans, Modular invariants, graphs and $\alpha$-induction for nets of subfactors, II. Commun. Math. Phys. **200**, 57–103 (1999)
9. J. Böckenhauer, D. Evans, Modular invariants, graphs and $\alpha$-induction for nets of subfactors, III. Commun. Math. Phys. **205**, 183–228 (1999)
10. K.-H. Rehren, Canonical tensor product subfactors. Commun. Math. Phys. **211**, 395–406 (2000)
11. J. Böckenhauer, D. Evans, Y. Kawahigashi, On $\alpha$-induction, chiral projectors and modular invariants for subfactors. Commun. Math. Phys. **208**, 429–487 (1999)
12. J. Böckenhauer, D. Evans, Y. Kawahigashi, Chiral structure of modular invariants for subfactors. Commun. Math. Phys. **210**, 733–784 (2000)
13. D. Evans, P. Pinto, Subfactor realizations of modular invariants. Commun. Math. Phys. **237**, 309–363 (2003)
14. A. Davydov, M. Müger, D. Nikshych, V. Ostrik, The Witt group of non-degenerate braided fusion categories. arXiv:1009.2117
15. F. Xu, Mirror extensions of local nets. Commun. Math. Phys. **270**, 835–847 (2007)
16. J. Fröhlich, J. Fuchs, I. Runkel, C. Schweigert, Correspondences of ribbon categories. Ann. Math. **199**, 192–329 (2006)
17. M. Bischoff, Y. Kawahigashi, R. Longo, Characterization of 2D rational local conformal nets and its boundary conditions: the maximal case. arXiv:1410.8848
18. L. Kong, I. Runkel, Algebraic structures in Euclidean and Minkowskian two-dimensional conformal field theory. arXiv:0902.3829
19. M. Izumi, The structure of sectors associated with Longo-Rehren inclusions. I. General theory. Commun. Math. Phys. **213**, 127–179 (2000)
20. K.-H. Rehren, Braid group statistics and their superselection rules, in *The Algebraic Theory of Superselection Sectors*, ed. by D. Kastler (World Scientific, Singapore, 1990), pp. 333–355
21. J. Fuchs, I. Runkel, C. Schweigert, TFT construction of RCFT correlators I: partition functions. Nucl. Phys. B **646**, 353–497 (2002)
22. J. Fuchs, I. Runkel, C. Schweigert, TFT construction of RCFT correlators, II. Nucl. Phys. B **678**, 511–637 (2004)

23. J. Fuchs, I. Runkel, C. Schweigert, TFT construction of RCFT correlators, III. Nucl. Phys. B **694**, 277–353 (2004)
24. J. Fuchs, I. Runkel, C. Schweigert, TFT construction of RCFT correlators, IV. Nucl. Phys. B **715**, 539–638 (2005)
25. L. Kong, I. Runkel, Morita classes of algebras in modular tensor categories. Adv. Math. **219**, 1548–1576 (2008)

# Chapter 5
# Applications in QFT

**Abstract** We review some applications of the abstract theory presented in the preceding chapters in the context of local quantum field theory. The methods developed in Chap. 4 prove to be most efficient to deal with QFT with boundaries, and to classify the boundary conditions.

We review some applications of the above abstract theory in the context of local quantum field theory. In a nutshell, Q-systems provide a complete characterization of (finite index) extensions of local quantum field theories, and the notions and operations discussed in the main body of this work have counterparts in conformal QFT that are of interest for the construction and classification of local extensions and of boundary conditions. More details can be found in [1–5].

The enormous benefit of the approach lies in the fact that, once the validity of the formalism is established, one does not need any dynamical details of the quantum field theory at hand, except the knowledge of its representation category as a braided C* tensor category. In turn, as Examples 3.1 and 4.17 show, this information typically requires very few data (like the fusion rules and the twist parameters $\kappa_\rho$) which in many cases uniquely fix the category.

## 5.1 Basics of Algebraic Quantum Field Theory

### 5.1.1 Local Nets

The additional feature in quantum field theory is the local structure: quantum fields are operator-valued distributions in spacetime, such that the support of the test function specifies the localization of field operators. In the algebraic approach [6] one rather considers local algebras $\mathscr{A}(O)$ of bounded operators generated by quantum fields evaluated on test functions with a given spacetime support $O$.

In fact, it is not necessary to assume that the local algebras are generated by actual quantum fields. It is sufficient to assume that the net of local algebras is isotonous, i.e., $O_1 \subset O_2$ implies $\mathscr{A}(O_1) \subset \mathscr{A}(O_2)$.

 
M. Bischoff et al., *Tensor Categories and Endomorphisms of von Neumann Algebras*, SpringerBriefs in Mathematical Physics, DOI 10.1007/978-3-319-14301-9_5

Thus, rather than with a single von Neumann algebra, one deals with a directed net of von Neumann algebras $\mathscr{A}(O)$, where $O$ runs over a suitable family of connected open regions in spacetime, and $\mathscr{A}(O)$ are the von Neumann algebras generated by local observables localized in $O$. (If $O$ has a sufficiently large causal complement, then the algebra $\mathscr{A}(O)$ does not depend on the representation in which the weak closure is taken.) The regions $O$ can be chosen to be doublecones (intersections of a future and a past lightcone)

in Minkowski spacetime $M^D = (\mathbb{R}^D, \eta)$, or intervals $I \subset \mathbb{R}$ where $\mathbb{R}$ is the "spacetime" of a chiral quantum field theory.

The use of $\mathbb{R}$ as the "spacetime" of a chiral quantum field theory is due to the feature of conformal quantum field theory in two dimensions that it necessarily contains fields (notably the stress-energy tensor which is the local generator of diffeomorphism covariance) that depend only on either lightcone coordinate $t + x$ or $t - x$. The net of local algebras generated by chiral fields is therefore indexed by the intervals $I \subset \mathbb{R}$. By virtue of the conformal symmetry, a conformal net on $\mathbb{R}$ actually extends to a net (more precisely: a pre-cosheaf) on $S^1$ by identifying $\mathbb{R}$ with $S^1$ minus a point; but this feature will not be essential for the applications that we are going to review.

In two spacetime dimensions, the spacelike complement of a doublecone has two connected components (wedges). In chiral theories, the spacelike complement of an interval is just its complement in $\mathbb{R}$, which is a disconnected union of two halfrays.

The quasilocal algebra $\mathscr{A}_{\mathrm{ql}}$ associated with a net $\mathscr{A} : O \mapsto \mathscr{A}(O)$ is the C* algebra defined as the inductive C* limit as $O$ exhausts the entire spacetime. A group $G$ of spacetime symmetries (Poincaré group, conformal group) is assumed to act on $\mathscr{A}_{\mathrm{ql}}$ as automorphisms $\alpha_g$ such that $\alpha_g(\mathscr{A}(O)) = \mathscr{A}(gO)$.

The principle of causality (locality) expresses the absence of superluminal causal influences. In quantum theory, it asserts that observables localized at spacelike distance must commute with each other. Thus, if two regions $O_1$, $O_2$ are spacelike separated (in the chiral case: disjoint), then $[\mathscr{A}(O_1), \mathscr{A}(O_2)] = \{0\}$ (as subalgebras of $\mathscr{A}_{\mathrm{ql}}$), or equivalently

$$\mathscr{A}(O) \subset \mathscr{A}(O')',$$

where $O'$ is the causal complement of $O$, and $\mathscr{A}(O')$ the C* algebra generated by $\mathscr{A}(\widehat{O})$, $\widehat{O} \subset O'$, and $\mathscr{A}(O')'$ its commutant in $\mathscr{A}_{\mathrm{ql}}$.

An overview of the consequences of these axioms (isotony, covariance, locality, vacuum representation) in chiral conformal QFT can be found in [7, 8].

There is a variety of methods to construct local conformal nets. Free field nets can be constructed as CAR or CCR algebras, equipped with a vacuum state. Local nets associated with affine Kac-Moody algebras can be constructed from unitary implementers of local gauge transformations acting as automorphisms of the CAR algebra, giving rise to projective representations of loop groups [9]. Local nets associated with a chiral stress-energy tensor can similarly be obtained from unitary implementers of local diffeomorphisms acting as automorphisms of the CAR algebra; an alternative, more explicit construction from a given heighest-weight representation of the Virasoro algebra is given in [8]. By orbifold (fixed point) and coset (relative commutants) constructions, one can construct new nets from given ones. Finally, the extension of local nets by commutative Q-systems will be described in Sect. 5.2.

### *5.1.2 Representations and DHR Endomorphisms*

Nets of local algebras possess inequivalent Hilbert space representations. A representation $\pi$ is covariant if the automorphisms $\alpha_g$ are implemented by a unitary representation $U_\pi(g)$ on the representation Hilbert space, $\pi(\alpha_g(a)) = U_\pi(g)\pi(a)U_\pi(g)^*$. A representation $\pi$ is said to have positive energy if the generator of the unitary one-parameter group $U_\pi(t)$ corresponding to the subgroup of time translations has positive spectrum. We assume that there is a unique vacuum representation $\pi_0$, i.e., a faithful positive-energy representation with an invariant ground state $\Omega$, $U(g)\Omega = \Omega$, and we assume that in the vacuum representation a stronger version of locality holds, namely Haag duality:

$$\pi_0(\mathscr{A}(O)) = \pi_0(\mathscr{A}(O'))'.$$

Under these standard assumptions, one can show that the local algebras $A(O)$ are infinite factors. Moreover [10], an important class of positive-energy representations (in two-dimensional conformal QFT: all positive-energy representations) can be described in terms of DHR endomorphisms $\rho$ of the quasilocal algebra $\mathscr{A}_{\text{ql}}$ such that

$$\pi = \pi_0 \circ \rho.$$

DHR endomorphisms are *localized* in some region $O$ in the sense that the restriction of $\rho$ to the algebra $\mathscr{A}(O')$ of the causal complement acts like the identity; and *transportable* in the sense that for every other region $\widehat{O}$, there is an endomorphism $\widehat{\rho}$ localized in $\widehat{O}$ which is unitarily equivalent, namely, there is a unitary **charge transporter** $u \in \mathscr{A}_{\text{ql}}$ (actually localized in any doublecone that contains $O$ and $\widehat{O}$) such that $\widehat{\rho} = \text{Ad}_u \circ \rho$:

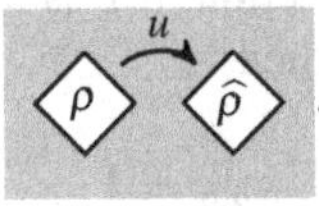

.

By Haag duality it follows that $\rho(\mathscr{A}(O)) \subset \mathscr{A}(O)$ if $\rho$ is localized in $O$, i.e., $\rho$ restricts to an endomorphism of the von Neumann algebra $N = \mathscr{A}(O)$.

The composition $\rho_1 \circ \rho_2$ of DHR endomorphisms is again a DHR endomorphism. Inertwiners between DHR endomorphisms are defined as operators $t \in \mathscr{A}_{\text{ql}}$ satisfying

$t\rho_1(a) = \rho_2(a)t$ for all $a \in \mathscr{A}_{\mathrm{ql}}$. By Haag duality, it follows that $t \in \mathscr{A}(O)$ if $\rho_i$ are localized in $O_i$ and $O_1 \cup O_2 \subset O$. In particular, all intertwiners among DHR endomorphisms localized in the same region $O$ are elements of $\mathscr{A}(O)$.

In this way, picking any fixed region $O$ and putting $N = \mathscr{A}(O)$, the restrictions of DHR endomorphisms localized in $O$ form a C* tensor subcategory of End($N$). One can show [7, Theorem 2.3] that this subcategory is full, i.e., every intertwiner between $\rho_1$ and $\rho_2$ regarded as endomorphisms of the von Neumann algebra $N$ is also an intertwiner between $\rho_1$ and $\rho_2$ regarded as endomorphisms of the C* algebra $\mathscr{A}_{\mathrm{ql}}$ (local intertwiners = global intertwiners). In particular, notions like sector, conjugates and dimension have the same meaning for DHR endomorphisms as endomorphisms of $\mathscr{A}_{\mathrm{ql}}$ and as endomorphisms of $N$.

We denote by $\mathscr{C}^{\mathrm{DHR}}(\mathscr{A})|_O$ the full subcategory of $\mathrm{End}_0(N)$, whose objects are the DHR endomorphisms of finite dimension, localized in $O$, and by $\mathscr{C}^{\mathrm{DHR}}(\mathscr{A})$ the C* tensor category of all DHR endomorphisms of finite dimension. Example 3.1 specifies the DHR category of the chiral Ising model.

### 5.1.3 DHR Braiding

The C* tensor category $\mathscr{C}^{\mathrm{DHR}}(\mathscr{A})$ is equipped with a distinguished unitary braiding $\varepsilon_{\rho,\sigma} \in \mathrm{Hom}(\rho\,\sigma, \sigma\,\rho)$. It is defined using unitary charge transporters $u_\rho \in \mathrm{Hom}(\rho, \widehat{\rho})$ and $u_\sigma \in \mathrm{Hom}(\sigma, \widehat{\sigma})$, such that $\widehat{\rho}$ is localized to the spacelike right (in the chiral case: in the future) of $\widehat{\sigma}$:

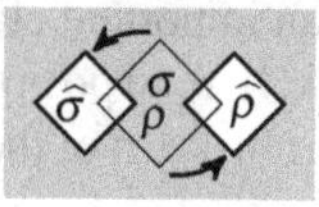
.

One shows with Haag duality that the auxiliary endomorphisms $\widehat{\rho}$ and $\widehat{\sigma}$, being localized at spacelike distance, commute with each other, and defines

$$\varepsilon_{\rho,\sigma} := (u_\sigma \times u_\rho)^* \circ (u_\rho \times u_\sigma) \equiv \sigma(u_\rho^*)u_\sigma^* u_\rho \rho(u_\sigma) \in \mathrm{Hom}(\rho\sigma, \sigma\rho). \quad (5.1.1)$$

This unitary does not depend on the choice of $\widehat{\rho}, \widehat{\sigma}$ with the specified relative localization, nor on the choice of the charge transporters $u_\rho$, $u_\sigma$. It satisfies the defining properties of a braiding. By construction, if $\rho$ is localized to the spacelike right (in the chiral case: in the future) of $\sigma$, then

$$\varepsilon_{\rho,\sigma} = 1,$$

because one may just choose $\widehat{\rho} = \rho, \widehat{\sigma} = \sigma$, and $u_\rho = u_\sigma = 1$. In contrast, if $\rho$ is localized to the spacelike left (past) of $\sigma$, one will have $\varepsilon_{\sigma,\rho} = 1$ but $\varepsilon_{\rho,\sigma} \neq 1$ in general, because the braiding $\varepsilon^+_{\rho,\sigma} \equiv \varepsilon_{\rho,\sigma}$ and its opposite $\varepsilon^-_{\rho,\sigma} \equiv \varepsilon^*_{\sigma,\rho}$ differ in low-dimensional QFT, due to the two connected components of the causal complement.

In four dimensional QFT, the braiding is degenerate: $\varepsilon_{\rho,\sigma}\varepsilon_{\sigma,\rho} = \mathbf{1}$, i.e., it is a permutation symmetry, and the twist parameter $\kappa_\rho = \pm 1$ distinguishes fermionic and bosonic sectors [10]. In chiral conformal QFT, the conformal spin-statistics theorem [7] relates the twist parameter $\kappa_\rho = e^{2\pi i h_\rho}$ of a sector to the lowest eigenvalue $h_\rho$ of $L_0$.

If both $\rho$ and $\sigma$ are localized in $O$, then $\varepsilon_{\rho,\sigma} \in \mathscr{A}(O)$, hence the DHR braiding restricts to each $\mathscr{C}^{\rm DHR}(\mathscr{A})|_O$. The local structure of a QFT net therefore provides us intrinsically with a braided C* tensor category, the arena of the abstract theory of the previous chapters.

Of particular interest in the context of the present work is the case when the quantum field theory $\mathscr{A}$ possesses only finitely many irreducible DHR sectors of finite dimension. In chiral conformal QFT, this property (referred to as "completely rationality"), is known to follow from the split property and Haag duality for intervals. Many models of interest, including the chiral Virasoro models with central charge $c < 1$, are completely rational. The case $c = \frac{1}{2}$ is the chiral Ising model, Examples 3.1 and 4.17.

(Complete rationality should be regarded, however, rather as a technically useful regularity condition with far-reaching consequences, than an axiom based on physical principles—since important models, like the $u(1)$ current algebra, do not share this property.)

In completely rational chiral models, the DHR braiding is non-degenerate [11], making the braided category $\mathscr{C}^{\rm DHR}(\mathscr{A})$ a modular category, cf. Sect. 4.11. Moreover, the global dimension of $\mathscr{C}^{\rm DHR}(\mathscr{A})$ (i.e., the quantity $\sum_{[\rho]\,{\rm irr}} \dim(\rho)^2$, Eq. (3.0.1)) coincides with the $\mu$-index of $\mathscr{A}$ which measures the violation of Haag duality for pairs of disconnected intervals [11, 12]. Thus, the presence of DHR sectors can be "detected" by inspection of the two-interval subfactor

$$\pi_0\big(\mathscr{A}(I_1 \cup I_2)\big) \subset \pi_0\big(\mathscr{A}((I_1 \cup I_2)')\big)'$$

where $I_1, I_2$ are any pair of non-touching intervals. Recall that the global dimension also is the common dimension of $\Theta$ in all irreducible full centre Q-systems by Proposition 4.43. In particular, the two-interval subfactor is isomorphic with the subfactor described by the canonical Q-system [1].

We now turn to the interpretations of Q-systems and the various operations on them, in the QFT context.

## 5.2 Local and Nonlocal Extensions

### *5.2.1 Q-Systems for Quantum Field Theories*

A Q-system in a C* tensor category $\mathscr{C} \subset \mathrm{End}_0(N)$ describes an extension $N \subset M$. A Q-system $\mathbf{A} = (\theta, w, x)$ in the category $\mathscr{C} = \mathscr{C}^{\rm DHR}(\mathscr{A})$ describes a family of extensions

$$\mathscr{A}(O) \subset \mathscr{B}(O)$$

in very much the same way. Namely, let $\mathscr{B}$ be the * algebra generated by $\mathscr{A}_{\rm ql}$ and $v$ subject to the relations

$$v \cdot a = \theta(a) \cdot v, \quad v^2 = x \cdot v, \quad v^* = w^* x^* \cdot v,$$

such that $\mathscr{B} = \mathscr{A}_{\rm ql} \cdot v$ as a vector space. Embed $\mathscr{A}_{\rm ql}$ by $\iota(a) = aw^* \cdot v$ as a * subalgebra. Define * subalgebras

$$\mathscr{B}(O) := \mathscr{A}(O) u \cdot v$$

where $u \in \mathscr{A}_{\rm ql}$ is a unitary such that $\widehat{\theta} = \mathrm{Ad}_u \circ \theta$ is localized in $O$. Because $(u \times u) \circ x \circ u^* \in \mathrm{Hom}(\widehat{\theta}, \widehat{\theta}^2) \subset \mathscr{A}(O)$ and $(u \times u) \circ x \circ w \in \mathrm{Hom}(\mathrm{id}, \widehat{\theta}^2) \subset \mathscr{A}(O)$, $\mathscr{B}(O)$ are indeed * algebras. In fact, $\mathscr{A}(O) \subset \mathscr{B}(O)$ is precisely the von Neumann algebra extension of $\mathscr{A}(O)$ by the Q-system $(\widehat{\theta} = \mathrm{Ad}_u \circ \theta, u \circ w, (u \times u) \circ x \circ u^*)$.

One obtains a net of von Neumann algebras $O \mapsto \mathscr{B}(O)$ extending $\mathscr{A}(O)$, and $\mathscr{B}$ is its inductive limit as the regions $O$ exhaust the entire spacetime.

Charged intertwiners $\psi_\rho$, defined for $\rho \prec \theta$ as in Remark 3.12, are elements of $\mathscr{B}(O)$ whenever $\rho$ is localized in $O$, and these operators together with $\mathscr{A}(O)$ generate $\mathscr{B}(O)$. As $O$ varies, these operators are the substitute of charged "fields" in the language of algebraic QFT.

The charged intertwiners create charged states from the vacuum as follows [1]. The positive map

$$\mu : b \mapsto d_{\mathbf{A}}^{-1} \cdot w^* \overline{\iota}(b) w$$

is a conditional expectation $\mu : \mathscr{B} \to \mathscr{A}$. It allows to extend the vacuum state $\omega_0$ on $\mathscr{A}$ to a vacuum state $\omega := \omega_0 \circ \mu$ on $\mathscr{B}$. Since $\mu(\psi_\rho \psi_\rho^*) \in \mathrm{Hom}(\rho, \rho)$ is a multiple of $\mathbf{1}$ if $\rho$ is irreducible, we may assume it to be $= \mathbf{1}$ by normalizing $\psi_\rho$. Then one has

$$\omega_0 \circ \mu(\psi_\rho \iota(a) \psi_\rho^*) = \omega_0(\rho(a) \mu(\psi_\rho \psi_\rho^*)) = \omega_0 \circ \rho(a).$$

Thus, in the GNS representation $\pi$ of the state $\omega$, the vector $\pi(\psi_\rho^*)\Omega_\omega$ belongs to the DHR representation $\pi_0 \circ \rho$ of $\mathscr{A}$. Indeed, upon restriction to $\mathscr{A}$, the GNS representation of $\omega$ is equivalent to the DHR representation $\pi_0 \circ \theta$ of $\mathscr{A}$.

The net $\mathscr{B}$ is by construction relatively local w.r.t. the subnet $\mathscr{A}$: if $b = auv \in \mathscr{B}(O)$ with $a \in \mathscr{A}(O)$, and $a' \in \mathscr{A}(O')$, then

$$b \cdot a' = auv \cdot a' = au\theta(a')v = a\widehat{\theta}(a')uv = aa'uv = a' \cdot auv = a' \cdot b,$$

where we have used the localization of $\widehat{\theta}$ and the local commutativity of $a$ with $a'$. In fact, every relatively local net of extensions of finite index arises this way [1].

An extension $\mathscr{B}$ of $\mathscr{A}$ is in general not local. It is local iff $u_1 v$ commutes with $u_2 v$ whenever $\theta_1 = \mathrm{Ad}_{u_1} \circ \theta$ and $\theta_2 = \mathrm{Ad}_{u_2} \circ \theta$ are localized in spacelike separated regions $O_1$, $O_2$. But

$$u_1 v \cdot u_2 v = u_1 \theta(u_2) x v, \quad u_2 v \cdot u_1 v = u_2 \theta(u_1) x v,$$

which are equal iff $\varepsilon_{\theta,\theta} \circ x = x$, by the definition of the DHR braiding Eq. (5.1.1). Thus, $\mathscr{B}$ is local iff the Q-system $(\theta, w, x)$ is commutative.

Given a local extension $\mathscr{A} \subset \mathscr{B}$, one may apply $\alpha$-induction (cf. Sect. 4.6) to the DHR endomorphisms $\rho$ of $\mathscr{A}$, defining endomorphisms $\alpha_\rho^\pm$ of $\mathscr{B}$. These are, however, in general not DHR endomorphisms of $\mathscr{B}$, since they act trivially only on one of the two components of the causal complement of the localization region of $\rho$. DHR endomorphisms of $\mathscr{B}$ are obtained as sub-endomorphisms which are contained in both $\alpha_\rho^+$ and $\alpha_\sigma^-$ for some $\rho, \sigma \in \mathscr{C}^{\mathrm{DHR}}(\mathscr{A})$. The common ("ambichiral") sub-endomorphisms are counted by the numbers $Z_{\rho,\sigma} = \dim \mathrm{Hom}(\alpha_\rho^-, \alpha_\sigma^+)$, cf. Eq. (4.6.3).

By classifying (commutative) Q-systems within the DHR category of a given completely rational quantum field theory, one obtains a classification of its (local) extensions. This program has been completed (profiting from existence and uniqueness results of [13] and the previous classifications of modular invariant matrices in [14]) for the local extensions of chiral nets associated with the stress-energy tensor with central charge $c < 1$, which are known to be completely rational [15]. All models in this classification can be realized by coset constructions, except one which arises as a mirror extension (cf. Sect. 4.7) of a coset extension. The classification of relatively local extensions with $c < 1$ (which is of interest in the presence of boundaries, Sect. 5.3) can be found in [16]; and the classification of local two-dimensional extensions (Sect. 5.2.2) with $c < 1$ was achieved in [17].

### 5.2.2 Two-Dimensional Extensions

The chiral observables of a two-dimensional conformal QFT are given by a tensor product of two chiral nets $\mathscr{A}_2 := \mathscr{A}_+ \otimes \mathscr{A}_-$ such that

$$\mathscr{A}_2(O) = \mathscr{A}_+(I) \times \mathscr{A}_-(J)$$

if

$$O = I \times J = \{(t,x) : t + x \in I, t - x \in J\} :$$ 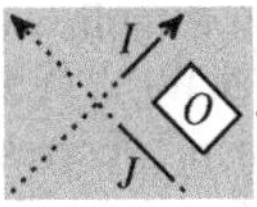

.

Its DHR endomorphisms are direct sums of $\rho_+ \otimes \rho_- \in \mathscr{C}^{\mathrm{DHR}}(\mathscr{A}_+) \otimes \mathscr{C}^{\mathrm{DHR}}(\mathscr{A}_-)$. From the definition of the DHR braidings, and because $O_1 = I_1 \times J_1$ is in the right spacelike complement of $O_2 = I_2 \times J_2$ if and only if $I_1$ is in the future of $I_2$ and $J_1$ is in the past of $J_2$, it follows that the braiding of $\mathscr{A}_2$ is given by $\varepsilon^+ \otimes \varepsilon^-$. Therefore, as a braided category, $\mathscr{C}^{\mathrm{DHR}}(\mathscr{A}_2) = \mathscr{C}^{\mathrm{DHR}}(\mathscr{A}_+) \boxtimes \mathscr{C}^{\mathrm{DHR}}(\mathscr{A}_-)^{\mathrm{opp}}$.

In particular, if the chiral nets $\mathscr{A}_+$ and $\mathscr{A}_-$ are isomorphic, then the canonical Q-system gives rise to a local two-dimensional extension $\mathscr{B}_2$ of $\mathscr{A}_2 = \mathscr{A} \otimes \mathscr{A}$, which is also known as the "Cardy type" extension. Its charged fields carry conjugate charges w.r.t. the $+$ and $-$ chiral observables. For the construction of this extension, it is actually not essential that $\mathscr{A}_+$ and $\mathscr{A}_-$ are isomorphic, but it is sufficient that they have isomorphic DHR categories. Obviously, one may as well construct a Cardy type extension based on any pair of isomorphic subcategories of $\mathscr{C}^{\mathrm{DHR}}(\mathscr{A}_+)$ and of $\mathscr{C}^{\mathrm{DHR}}(\mathscr{A}_-)$.

A more general class of local two-dimensional extensions of $\mathscr{A}_2$ was constructed in [18], by exhibiting the numerical coefficients of the Q-system $(\Theta, W, X)$ in $\mathscr{C}^{\mathrm{DHR}}(\mathscr{A}) \boxtimes \mathscr{C}^{\mathrm{DHR}}(\mathscr{A})^{\mathrm{opp}}$ by a method involving chiral $\alpha$-induction along a possibly noncommutative chiral Q-system (the "$\alpha$-induction construction"). The multiplicities of the irreducible subsectors $\rho \otimes \overline{\sigma} \prec \Theta$ coincide with the matrix elements $Z_{\rho,\sigma} = \dim \mathrm{Hom}(\alpha^-_\rho, \alpha^+_\sigma)$ of the modular invariants, mentioned before, cf. Sect. 5.2.1.

### 5.2.3 Left and Right Centre

In general, the extension $\mathscr{A} \subset \mathscr{B}$ described by a Q-system $\mathbf{A}$ in $\mathscr{C}^{\mathrm{DHR}}(\mathscr{A})$ will be nonlocal. Since the left and right centres $C^\pm[\mathbf{A}]$ of $\mathbf{A}$ are commutative Q-systems, cf. Sect. 4.8, they correspond to local extensions $\mathscr{B}^\pm_{\mathrm{loc}}$ intermediate between $\mathscr{A}$ and $\mathscr{B}$.

In [4], we have identified these local intermediate extensions with relative commutants

$$\mathscr{B}^+_{\mathrm{loc}}(O) := \mathscr{B}(W_L)' \cap \mathscr{B}(W'_R), \quad \text{resp.} \quad \mathscr{B}^-_{\mathrm{loc}}(O) := \mathscr{B}(W_R)' \cap \mathscr{B}(W'_L).$$

Here, the wedges $W_L$ and $W_R$ are the left and right components of the spacelike complement of the doublecone $O$ (resp. the past and future complements of an interval in the chiral case):

,
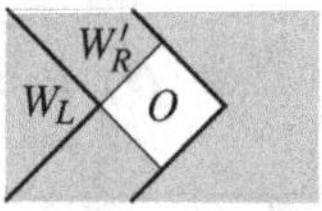

.

In order to establish this result, one has to verify that the relative commutant $\mathscr{B}^+_{\mathrm{loc}}(O) = \mathscr{B}(W_L)' \cap \mathscr{B}(W'_R)$ is intermediate between $\mathscr{A}(O) \subset \mathscr{B}(O)$, and that the projection $p^+_{\mathrm{loc}}$ corresponding to the intermediate extension coincides with the right centre projection $p^+$ of the Q-system for $\mathscr{A}(O) \subset \mathscr{B}(W_L)' \cap \mathscr{B}(W'_R) \subset \mathscr{B}(O)$. Thanks to Proposition 4.26, it is sufficient to prove that $p^+_{\mathrm{loc}}$ satisfies the relation Eq. (4.8.1), and that $\mathscr{B}^+_{\mathrm{loc}}(O)$ is maximal with this property.

The inclusion $\mathscr{A}(O) \subset \mathscr{B}^+_{\mathrm{loc}}(O)$ is obvious by isotony of $\mathscr{B}$ and relative locality of $\mathscr{B}$ w.r.t. $\mathscr{A}$. The inclusion $\mathscr{B}^+_{\mathrm{loc}}(O) \subset \mathscr{B}(O)$ can be established with the help of Haag duality for wedges, which was assumed to be valid for the net $\mathscr{A}$. The intersection $\mathscr{B}^+_{\mathrm{loc}}(O)$ is therefore the maximal subalgebra of $\mathscr{B}(O)$ commuting with $\mathscr{B}(W_L)$. One has $\mathscr{B}^+_{\mathrm{loc}}(O) = \mathscr{A}(O)v^+_{\mathrm{loc}} = \mathscr{A}(O)p^+_{\mathrm{loc}}v$. This algebra commutes with $\mathcal{B}(W_L)$ iff the generator $p^+_{\mathrm{loc}}v$ of $\mathcal{B}^+_{\mathrm{loc}}(O)$ commutes with the generator $\widehat{v} = uv$ of $\mathcal{B}(W_L)$, where $u \in \mathrm{Hom}(\theta, \widehat{\theta})$ is a unitary charge transporter taking $\theta$ to $\widehat{\theta}$ localized in $W_L$. Now, $p^+_{\mathrm{loc}}v \cdot uv = p^+_{\mathrm{loc}}\theta(u)xv = \theta(u)p^+_{\mathrm{loc}}xv$, whereas $uv \cdot p^+_{\mathrm{loc}}v = u\theta(p^+_{\mathrm{loc}})xv$. Commutativity is therefore equivalent to $p^+_{\mathrm{loc}}x = \theta(u)^*u\theta(p^+_{\mathrm{loc}})x$, which is Eq. (4.8.1) by the definition of the DHR braiding. The claim, $p^+_{\mathrm{loc}} = p^+$, then follows by the maximality of $\mathcal{B}^+_{\mathrm{loc}}(O)$ and the characterization of $p^+$ in Proposition 4.26.

In establishing this result, it is again essential that the braiding is the DHR braiding Eq. (5.1.1), defined in terms of unitary charge transporters. This explains why a similar interpretation of the left and right centre as the Q-system of some relative commutant cannot be given in a general braided subcategory of $\mathrm{End}_0(N)$ for a single von Neumann algebra $N$. It would be interesting to have such a theory, which would require—in addition to a braided tensor category $\mathcal{C} \subset \mathrm{End}_0(N)$—as additional data a splitting of $N'$ into two commuting subalgebras $N' = N_L \vee N_R$, and unitary intertwiners between endomorphisms $\rho \in \mathrm{End}_0(N)$ and endomorphisms of $N_L$ and of $N_R$, connected to the given braiding by (versions of) Eq. (5.1.1).

### 5.2.4 *Braided Product of Extensions*

According to Lemma 4.29, the braided products $\mathbf{A} \times^\pm \mathbf{B}$ of two Q-systems $\mathbf{A} = (\theta^{\mathbf{A}}, w^{\mathbf{A}}, x^{\mathbf{A}})$ and $\mathbf{B} = (\theta^{\mathbf{B}}, w^{\mathbf{B}}, x^{\mathbf{B}})$ describe extensions $M^\pm$ which are generated by the algebra $N$ and the generators $v^{\mathbf{A}}$ and $v^{\mathbf{B}}$ such that $Nv^{\mathbf{A}} = M^{\mathbf{A}}$ and $Nv^{\mathbf{B}} = M^{\mathbf{B}}$ are intermediate algebras, and the generators $v^{\mathbf{A}}, v^{\mathbf{B}}$ satisfy the commutation relations

$$v^{\mathbf{B}}v^{\mathbf{A}} = \iota(\varepsilon^\pm_{\theta^{\mathbf{A}},\theta^{\mathbf{B}}}) \cdot v^{\mathbf{B}}v^{\mathbf{A}}. \tag{5.2.1}$$

These properties uniquely specify $M^\pm$.

The same holds true in the QFT setting, that is, the braided product $\mathcal{B}^\pm$ of two extensions $\mathcal{B}^{\mathbf{A}}$ and $\mathcal{B}^{\mathbf{B}}$ of a net $\mathcal{A}$ is generated by $\mathcal{A}$ and the generators $v^{\mathbf{A}}$, $v^{\mathbf{B}}$ subject to the same commutation relation Eq. (5.2.1).

It follows that for any unitary charge transporters $u_1 \in \mathrm{Hom}(\theta^{\mathbf{A}}, \widehat{\theta}^{\mathbf{A}})$, $u_2 \in \mathrm{Hom}(\theta^{\mathbf{B}}, \widehat{\theta}^{\mathbf{B}})$,

$$(u_2v^{\mathbf{B}})(u_1v^{\mathbf{A}}) = \iota(\varepsilon^\pm_{\widehat{\theta}^{\mathbf{A}},\widehat{\theta}^{\mathbf{B}}}) \cdot (u_1v^{\mathbf{A}})(u_2v^{\mathbf{B}}).$$

Now, if $\widehat{\theta}^{\mathbf{A}}$ is localized to the right (left) of $\widehat{\theta}^{\mathbf{B}}$, then $\varepsilon^+_{\widehat{\theta}^{\mathbf{A}},\widehat{\theta}^{\mathbf{B}}} = 1$ ($\varepsilon^-_{\widehat{\theta}^{\mathbf{A}},\widehat{\theta}^{\mathbf{B}}} = 1$), hence in these cases the generators $u_1v^{\mathbf{A}}$ and $u_2v^{\mathbf{B}}$ commute. Since the local algebras of an extension are generated by $\mathcal{A}(O)$ and $uv$ such that $\widehat{\theta} = \mathrm{Ad}_u \circ \theta$ is localized in $O$, it follows (using relative locality w.r.t. $\mathcal{A}$) that, as subalgebras of $\mathcal{B}^+$ resp. of $\mathcal{B}^-$, the algebras $\mathcal{B}^{\mathbf{A}}(O_1)$ and $\mathcal{B}^{\mathbf{B}}(O_2)$ commute with each other if $O_1$ is located to the spacelike right resp. left of $O_2$—but in general not in the converse order.

We paraphrase these situations by saying that the net $\mathscr{B}^{\mathbf{A}}$ is "right local" resp. "left local" w.r.t. the net $\mathscr{B}^{\mathbf{B}}$. Thus, the braided products of of two extensions $\mathscr{B}^{\mathbf{A}}$ and $\mathscr{B}^{\mathbf{B}}$ can be regarded as a quotient of the free product by the relations that identify the common subnets $\mathscr{A} \subset \mathscr{B}^{\mathbf{A}}$ and $\mathscr{A} \subset \mathscr{B}^{\mathbf{B}}$, and by the relations expressing that $\mathscr{B}^{\mathbf{A}}$ is "right local" resp. "left local" w.r.t. $\mathscr{B}^{\mathbf{B}}$.

(The same is true in the chiral case, replacing "right" by "future" and "left" by "past".)

Let $\mathscr{A}$ be a chiral QFT and $\mathbf{A}$ a Q-system in $\mathscr{C}^{\mathrm{DHR}}(\mathscr{A})$, describing a (nonlocal) chiral extension $\mathscr{B}$. The full centre $Z[\mathbf{A}]$ is a Q-system in $\mathscr{C}^{\mathrm{DHR}}(\mathscr{A} \otimes \mathscr{A})$, i.e., it describes a two-dimensional extension $\mathscr{B}_2$ of $\mathscr{A} \otimes \mathscr{A}$. Because the full centre is the right centre of the braided product $(\mathbf{A} \otimes 1) \times^+ \mathbf{R}$, we recognize the corresponding extension $\mathscr{B}_2$ as the relative commutant of right wedges of the nonlocal extension, obtained by the right-local braided product of the possibly nonlocal chiral extension $\mathscr{B} \otimes 1$ with the local canonical extension $\mathscr{B}_2^{\mathbf{R}}$.

By Proposition 4.33, the full centre coincides with the $\alpha$-induction construction which was originally found as a construction of two-dimensional local conformal QFT models out of chiral data. This result therefore not only gives a more satisfactory, purely algebraic interpretation of the $\alpha$-induction construction in terms of braided products of nets and relative commutants of wedge algebras, cf. Sect. 5.2.3; it also explains the fact (known before) that the latter depends only on the Morita equivalence class of the chiral Q-system in $\mathscr{C}$ [19]; namely two Q-systems in a modular tensor category $\mathscr{C}$ have the same full centre if and only if they are Morita equivalent [20].

Since moreover, every irreducible extension $\mathscr{B}_2$ of $\mathscr{A}_2$ is intermediate between $\mathscr{A}_2$ and an $\alpha$-induction extension [2, 3, 19], it follows that full centre extensions are precisely the maximal irreducible extensions (if the underlying chiral theory $\mathscr{A}$ is completely rational).

## 5.3 Hard Boundaries

A conformal quantum field theory with a "hard boundary" arises, when Minkowski spacetime $M^2$ is restricted to a halfspace, say the right halfspace $M^2_R = \{(t,x) : x > 0\}$. The stress-energy tensor defined on $M^2_R$ still splits into two chiral components, but if one imposes conservation of energy at the boundary, the two components are no longer independent fields, but instead they *coincide* as operator-valued distributions on $\mathbb{R}$ [19]. Thus, for $I$ and $J$ intervals such that $O = I \times J = \{(t,x) : t + x \in I, t - x \in J\}$ lies inside $M^2_R$ ($\Leftrightarrow I > J$ elementwise as subsets of $\mathbb{R}$), the local algebra of chiral observables is

$$\mathscr{A}_R(O) = \mathscr{A}(I) \vee \mathscr{A}(J)$$

rather than the tensor product $\mathscr{A}_+(I) \otimes \mathscr{A}_-(J)$. Here, the chiral algebras are generated by the stress-energy tensor and possibly further chiral fields whose boundary conditions might also impose an identification of the fields.

An analysis of local extensions $\mathscr{A}_R(O) \subset \mathscr{B}_R(O)$ on the halfspace was given in [19]. One finds (assuming $\mathscr{A}$ to be completely rational) that the local algebras of every maximal such extension are of the form

$$\mathscr{B}_R(O) = \mathscr{B}(K)' \cap \mathscr{B}(L) \qquad (O = I \times J \subset M^2_R) \tag{5.3.1}$$

where $\mathscr{A} \subset \mathscr{B}$ is a possibly nonlocal chiral extension (given by a Q-system in $\mathscr{C}^{\mathrm{DHR}}(\mathscr{A})$), and $K \subset L$ is the unique pair of open intervals such that $I \cup J = L \setminus \overline{K}$:

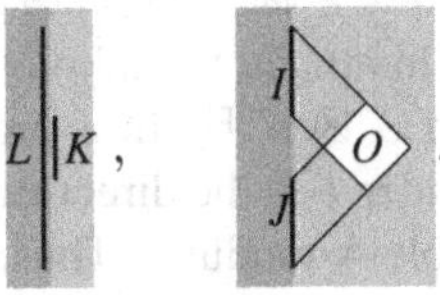


This formula is "holographic" in the sense that the local observables in a region $O \subset M^2_R$ are given in terms of operators in a chiral net that can be thought of as a net on the boundary.

The simplest case is the trivial chiral extension $\mathscr{B} = \mathscr{A}$. In this case, $\mathscr{B}_R(O)$ is generated by $\mathscr{A}_R(O) = \mathscr{A}(I) \vee \mathscr{A}(J)$ and charge transporters in $\mathscr{A}(L)$, namely unitary intertwiners transporting a DHR endomorphisms localized in $J$ to an equivalent DHR endomorphism localized in $J$. Accordingly, the charged generators for the subfactor $\mathscr{A}_R(O) \subset \mathscr{B}_R(O)$ "carry a charge $\rho$ in $I$ and a charge $\overline{\rho}$ in $J$". Indeed, under the split isomorphism between the von Neumann algebras $\mathscr{A}(I) \vee \mathscr{A}(J)$ and $\mathscr{A}(I) \otimes \mathscr{A}(J)$, the subfactor turns out to be isomorphic with the subfactor associated with the canonical Q-system Proposition 3.19 with $[\Theta_{\mathbf{R}}] = \bigoplus_{[\rho]\,\mathrm{irr}} \rho \otimes \overline{\rho}$.

For general chiral extensions $\mathscr{A} \subset \mathscr{B}$ with irreducible Q-system $\mathbf{A}$, the local subfactor $\mathscr{A}_R(O) \subset \mathscr{B}_R(O)$ for any bounded doublecone $O \subset M^2_R$ not touching the boundary is isomorphic to the subfactor obtained from the full centre Q-system $Z[\mathbf{A}]$, and hence depends (up to isomorphism) only on the Morita equivalence class (cf. Sect. 3.5) of the chiral Q-system $\mathbf{A}$.

As mentioned in Sect. 5.2.4, the full centre gives also a local net on the full two-dimensional Minkowski spacetime as an extension $\mathscr{B}_2 \supset \mathscr{A} \otimes \mathscr{A}$ of the tensor product of a pair of isomorphic chiral nets. Indeed, this net can be recovered from the maximal boundary net $\mathscr{B}_R$ by a procedure called "removing the boundary". It proceeds by taking the limit of a sequence of states on right wedge algebras $\mathscr{B}_R(W_R + a)$ as $a \in W_R$ tends to infinity ("far away from the boundary"). The net $\mathscr{B}_2$ can then be defined in the GNS Hilbert space of this state, which carries two commuting unitary representations of the Mbius group. First defining the local algebra $\mathscr{B}_2(W_R)$ of a single right wedge, and $\mathscr{B}_2(W'_R) := \mathscr{B}_s(W_R)'$ as its commutant, the two unitary representations of the translations are used to define the local algebras for general wedge regions, and the local algebras for doublecones by intersections of algebra for wedges.

The converse procedure of "adding a boundary" can also be performed algebraically [3]. Starting from an extension $\mathscr{A}_2 \subset \mathscr{B}_2$ defined on Minkowski spacetime, one can redefine the representation of its restriction to $M^2_R$, obtaining a reducible representation. Its decomposition yields a direct sum of boundary extensions $\mathscr{A}_R \subset \mathscr{B}_R$ related to chiral extensions $\mathscr{A} \subset \mathscr{B}$ by the "holographic formula" Eq. (5.3.1), which all give back $\mathscr{A}_2 \subset \mathscr{B}_2$ when the "boundary is removed". In particular, for every right wedge $W_R \subset M^2_R$ not touching the boundary, the subnets $\mathscr{A}_R(O) \subset \mathscr{B}_R(O)$ indexed by $O \subset W_R$ are all isomorphic to the subnets $\mathscr{A}_2(O) \subset \mathscr{B}_2(O)$, so that the boundary nets in the decomposition can be interpreted as different boundary conditions imposed on the original net $\mathscr{A}_2 \subset \mathscr{B}_2$.

The procedure of "adding a boundary" amounts, in the language of C*-tensor categories, to the tensor functor $T : \mathscr{C} \boxtimes \mathscr{C}^{\mathrm{opp}}, \rho \otimes \sigma \mapsto \rho\,\sigma$, taking Q-systems in $\mathscr{C} \boxtimes \mathscr{C}^{\mathrm{opp}}$ to Q-systems in $\mathscr{C}$. This functor is adjoint [21] to the full centre, taking Q-systems in $\mathscr{C}$ to Q-systems in $\mathscr{C} \boxtimes \mathscr{C}^{\mathrm{opp}}$. In is proven in [20] that the image of the full centre Q-system $Z[\mathbf{A}]$ under $T$ is the direct sum (in the sense of Sect. 4.2) of Q-systems given by the irreducible $\mathbf{A}$-modules. Thus, the hard boundary conditions are classified in 1:1 correspondence with the irreducible modules of the underlying chiral Q-system $\mathbf{A}$.

## 5.4 Transparent Boundaries

Whereas a hard boundary describing a QFT on a halfspace identifies the left- and right-moving chiral observables in the halfspace, a transparent boundary separates two possibly different quantum field theories $\mathscr{B}^L$ and $\mathscr{B}^R$ in the halfspaces $M^2_L, M^2_R$ on either side of the boundary:

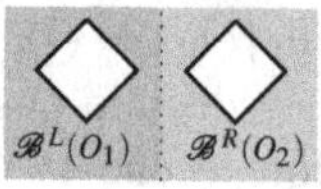


$$(O_1 \subset M^2_L, O_2 \subset M^2_R).$$

Physically speaking, the boundary is thought to separate regions with different dynamics, e.g., two different phases of a relativistic system with a phase transition. For the example of the Ising model, cf. [22] and Example 5.1.

The two theories are defined on the same Hilbert space, and share a tensor product $\mathscr{A}_+ \otimes \mathscr{A}_-$ of common chiral subtheories. The latter property arises from the physical assumption that energy *and* momentum are conserved at the boundary, which identifies the chiral stress-energy tensors on either side of the boundary [5].

Because the presence of the boundary cannot violate the principle of causality, quantum observables of $\mathscr{B}^L$ localized in the left halfspace $M^2_L$ must commute with observables of $\mathscr{B}^R$ localized in the right halfspace $M^2_R$ at spacelike separation.

Because the stress-energy tensor is the local generator of diffeomorphisms, the common chiral subtheory $\mathscr{A}_+ \otimes \mathscr{A}_-$ can be used to extend both theories to the full Minkowski spacetime.

Motivated by these two (heuristic) observations, one should define a transparent boundary as a pair of quantum field theories on two-dimensional Minkowski spacetime, sharing a common chiral subtheory, such that $\mathscr{B}^L$ is left-local w.r.t. $\mathscr{B}^R$. As we have seen before, such a pair is described by the braided product of two extensions of the common chiral subtheory, and every irreducible such pair is a quotient of the braided product.

The mathematical issue is therefore the central decomposition of the braided product of a pair of commutative Q-systems in $\mathscr{C}^{\mathrm{DHR}}(\mathscr{A}_+)\boxtimes\mathscr{C}^{\mathrm{DHR}}(\mathscr{A}_-)$. The centre of the braided product of extensions is given by Proposition 4.31 as a linear space isomorphic to $\mathrm{Hom}(\Theta^L,\Theta^R)$. In order to know its central projections, it must be computed as an algebra. This is precisely what we have achieved in Theorem 4.42, provided $\mathscr{C}^{\mathrm{DHR}}(\mathscr{A}_+)$ and $\mathscr{C}^{\mathrm{DHR}}(\mathscr{A}_-)$ are isomorphic as modular braided categories, and the pair of commutative Q-systems are full centres of chiral Q-systems $\mathbf{A}^L$ and $\mathbf{A}^R$. Namely, Theorem 4.42 classifies the transparent boundary conditions in 1:1 correspondence with the irreducible chiral $\mathbf{A}^L$-$\mathbf{A}^R$-bimodules.

In [5], this classification is further elaborated. As discussed in Sect. 4.12, the space $\mathrm{Hom}(\Theta^L,\Theta^R)$ has two distinguished bases, orthogonal w.r.t. the inner product Eq. (4.12.7): one arising by "diagonalizing" the left and right compositions with $\mathrm{Hom}(\Theta^L,\Theta^L)$ and $\mathrm{Hom}(\Theta^R,\Theta^R)$, the other corresponding to the minimal central projections of the braided product, i.e., the minimal projections in $\mathrm{Hom}(\Theta^L,\Theta^R)$ w.r.t. the convolution product Eq. (4.12.2). The unitary transition matrix is a generalized Verlinde matrix, and can be computed by its distinguishing property that it "diagonalizes" the bimodule fusion rules. Its matrix elements finally turn out to determine the specific identifications between charged fields of $\mathscr{B}^L$ and charged fields of $\mathscr{B}^R$, that make up the specific boundary conditions.

*Example 5.1* The special case where both Q-systems are the canonical one, i.e., the boundary between two conformal quantum field theories both isomorphic to the Cardy extension, has been given in Example 4.48. For the Ising model (i.e., the chiral net is given by the Virasoro net with central charge $c = 1$), one obtains three boundary conditions given by the three sets of linear dependencies between the charged generators $\Psi_{\sigma\otimes\sigma}$, $\Psi_{\tau\otimes\tau}$:

$$\text{(i)}\quad \Psi^L_{\tau\otimes\tau} = \Psi^R_{\tau\otimes\tau},\quad \Psi^L_{\sigma\otimes\sigma} = \Psi^R_{\sigma\otimes\sigma};$$

$$\text{(ii)}\quad \Psi^L_{\tau\otimes\tau} = \Psi^R_{\tau\otimes\tau},\quad \Psi^L_{\sigma\otimes\sigma} = -\Psi^R_{\sigma\otimes\sigma};$$

$$\text{(iii)}\quad \Psi^L_{\tau\otimes\tau} = -\Psi^R_{\tau\otimes\tau}.$$

The first case is the trivial boundary; the second the "fermionic" boundary where the field $\Psi_{\sigma\otimes\sigma}$ changes sign, and the third the "dual" boundary, in which there are two independent fields $\Psi^R_{\sigma\otimes\sigma}$ and $\Psi^L_{\sigma\otimes\sigma}$ (corresponding to the order and disorder parameter $\sigma$ and $\mu$ in [22]).

## 5.5 Further Directions

We have outlined the remarkably tight links between the abstract theory of Q-systems in braided C* tensor categories and the representation theory of conformal quantum field theories in two dimensions. Notably the classifications of "hard" and "transparent" boundary conditions have very natural counterparts in the abstract setting.

Thinking of systems with several transparent boundaries, some immediate questions arise: the juxtaposition of two boundaries is described by the (associative) braided product of three Q-systems. The individual boundary conditions are classified as **A**-**B**-bimodules and as **B**-**C**-bimodules. Thus, it is expected that the juxtaposition of boundary conditions is described in terms of the bimodule tensor product.

It is much less clear which mathematical structure should be expected to describe situations where two transparent boundaries *intersect* each other.

Finally, hard and transparent boundaries are only two "opposite extremes" in a wide spectrum of possible behaviour of chiral fields at a boundary [5]. It would be rewarding to describe also more general boundaries in terms of the present unifying framework.

## References

1. R. Longo, K.-H. Rehren, Nets of subfactors. Rev. Math. Phys. **7**, 567–597 (1995)
2. R. Longo, K.-H. Rehren, How to remove the boundary in CFT—an operator algebraic procedure. Commun. Math. Phys. **285**, 1165–1182 (2009)
3. S. Carpi, Y. Kawahigashi, R. Longo, How to add a boundary condition. Commun. Math. Phys. **322**, 149–166 (2013)
4. M. Bischoff, Y. Kawahigashi, R. Longo, Characterization of 2D rational local conformal nets and its boundary conditions: the maximal case. arXiv:1410.8848
5. M. Bischoff, Y. Kawahigashi, R. Longo, K.-H. Rehren, Phase boundaries in algebraic conformal QFT. arXiv:1405.7863
6. R. Haag, *Local Quantum Physics* (Springer, Berlin, 1996)
7. D. Guido, R. Longo, The conformal spin and statistics theorem. Commun. Math. Phys. **181**, 11–35 (1996)
8. S. Carpi, Y. Kawahigashi, R. Longo, Structure and classification of superconformal nets. Ann. Henri Poincaré **9**, 1069–1121 (2008)
9. A. Pressley, I. Segal, *Loop Groups* (Oxford University Press, Oxford, 1986)
10. S. Doplicher, R. Haag, J.E. Roberts, Local observables and particle statistics. I. Commun. Math. Phys. **23**, 199–230 (1971)
11. Y. Kawahigashi, R. Longo, M. Müger, Multi-interval subfactors and modularity of representations in conformal field theory. Commun. Math. Phys. **219**, 631–669 (2001)
12. R. Longo, F. Xu, Topological sectors and a dichotomy in conformal field theory. Commun. Math. Phys. **251**, 321–364 (2004)
13. A. Kirillov Jr., V. Ostrik, On $q$-analog of McKay correspondence and ADE classification of $sl(2)$ conformal field theories. Adv. Math. **171**, 183–227 (2002)
14. A. Cappelli, C. Itzykson, J.-B. Zuber, The $A$-$D$-$E$ classification of minimal and $A_1^{(1)}$ conformal invariant theories. Commun. Math. Phys. **113**, 1–26 (1987)

15. Y. Kawahigashi, R. Longo, Classification of local conformal nets. Case $c < 1$. Ann. Math. **160**, 493–522 (2004)
16. Y. Kawahigashi, R. Longo, U. Pennig, K.-H. Rehren, The classification of non-local chiral CFT with $c < 1$. Commun. Math. Phys. **271**, 375–385 (2007)
17. Y. Kawahigashi, R. Longo, Classification of two-dimensional local conformal nets with $c < 1$ and 2-cohomology vanishing for tensor categories. Commun. Math. Phys. **244**, 63–97 (2004)
18. K.-H. Rehren, Canonical tensor product subfactors. Commun. Math. Phys. **211**, 395–406 (2000)
19. R. Longo, K.-H. Rehren, Local fields in boundary CFT. Rev. Math. Phys. **16**, 909–960 (2004)
20. L. Kong, I. Runkel, Morita classes of algebras in modular tensor categories. Adv. Math. **219**, 1548–1576 (2008)
21. L. Kong, I. Runkel, Cardy algebras and sewing constraints, I. Commun. Math. Phys. **292**, 871–912 (2009)
22. B. Schroer, T.T. Truong, The order/disorder quantum field operators associated with the two-dimensional Ising model in the continuum limit. Nucl. Phys. B **144**, 80–122 (1978)

# Chapter 6
# Conclusions

Q-systems are a tool to describe extensions $N \subset M$ of an infinite von Neumann factor $N$ in terms of "data" referring only to $N$. We have extended this notion, well-known for subfactors, to the case when $M$ is admitted to be a finite direct sum of factors. Modules and bimodules of Q-systems are equivalent to homomorphisms between extensions. Decompositions of Q-systems and other operations defined in *braided* C* tensor categories: the centres, braided products and the full centre—which are known in the setting of abstract tensor categories—are interpreted in terms of the associated extensions of von Neumann algebras.

The meaning of these operations in the context of local quantum field theory is elaborated in [1]. Especially the determination of the centre of the von Neumann algebra which arises as the braided product of two commutative extensions, is a problem motivated by these applications. We have completely solved this task for the braided product of two full centres in *modular* C* tensor categories.

In the last section, we have given a brief outline of this and other applications of the theory of braided and modular C* tensor categories in the context of quantum field theory. It is here, where the interpretation in terms of endomorphisms of von Neumann algebras is most substantial, since local quantum observables are (selfadjoint) elements of von Neumann algebras. This application was not only our original motivation for the analysis presented in the main body of this work; it is also not an exaggeration to say that the (rather natural) appearance of *modular* C* categories in chiral conformal QFT, as an offspring of the original DHR theory designed for massive QFT in four spacetime dimensions, has triggered many of the developments described in this work.



M. Bischoff et al., *Tensor Categories and Endomorphisms of von Neumann Algebras*,
SpringerBriefs in Mathematical Physics, DOI 10.1007/978-3-319-14301-9_6

# Reference

1. M. Bischoff, Y. Kawahigashi, R. Longo, K.-H. Rehren, Phase boundaries in algebraic conformal QFT. arXiv:1405.7863